Comparing U.S. Army Systems with Foreign Counterparts

Identifying Possible Capability Gaps and Insights from Other Armies

John Gordon IV, John Matsumura, Anthony Atler, Scott Boston, Matthew E. Boyer, Natasha Lander, Todd Nichols

Prepared for the United States Army
Approved for public release; distribution unlimited

RAND ARROYO CENTER

For more information on this publication, visit www.rand.org/t/rr716

Library of Congress Control Number: 2015939204
ISBN: 978-0-8330-8721-8

Published by the RAND Corporation, Santa Monica, Calif.

© Copyright 2015 RAND Corporation

RAND® is a registered trademark.

Limited Print and Electronic Distribution Rights

This document and trademark(s) contained herein are protected by law. This representation of RAND intellectual property is provided for noncommercial use only. Unauthorized posting of this publication online is prohibited. Permission is given to duplicate this document for personal use only, as long as it is unaltered and complete. Permission is required from RAND to reproduce, or reuse in another form, any of its research documents for commercial use. For information on reprint and linking permissions, please visit www.rand.org/pubs/permissions.html.

The RAND Corporation is a research organization that develops solutions to public policy challenges to help make communities throughout the world safer and more secure, healthier and more prosperous. RAND is nonprofit, nonpartisan, and committed to the public interest.

RAND's publications do not necessarily reflect the opinions of its research clients and sponsors.

Support RAND

Make a tax-deductible charitable contribution at
www.rand.org/giving/contribute

www.rand.org

Preface

This fiscal year 2013 project was conducted on behalf of the Deputy Chief of Staff, Force Development (G-8) in Headquarters, Department of the Army. The project's title, *Comparing U.S. Army Systems with Foreign Counterparts: Identifying Possible Capability Gaps and Insights from Other Armies*, demonstrates the focus of this effort: to compare selected U.S. Army programs with their counterparts in a number of other armies around the world. As an organizing principle, the Army's warfighting functions were selected as a way of bounding and focusing the research. Warfighting functions include movement and maneuver, intelligence, fires, sustainment, mission command, and protection. During the course of the project, various areas of particular interest were selected in conjunction with the needs of the sponsor in order to better focus the research. For example, within the warfighting function of movement and maneuver, it was decided to concentrate on examining armored fighting vehicles and helicopters.

Given the very broad range of topics that had to be covered in this project, a decision was made in conjunction with the sponsor to conduct an overview of selected foreign systems and to focus on unclassified sources. Classified sources were periodically consulted during the research, and, when appropriate, classified insights were directly provided to the sponsor. Importantly, this report is based entirely on unclassified, open-source information. A number of foreign armies were selected for the comparisons. In some cases those armies are U.S. allies, while in other cases the army used for the comparison is a potential future competitor. Although the research was primarily focused on comparing the capabilities of material systems, such as armored fighting vehicles, logistics systems, and helicopters, within a warfighting function, crosscutting insights were developed where possible. Additionally, although the research focuses on material systems, where possible other implications for the U.S. Army in the areas of doctrine, training, and leader development were observed and highlighted.

This research was sponsored by the Director of Force Management and conducted within the RAND Arroyo Center's Force Development and Technology Program. RAND Arroyo Center, part of the RAND Corporation, is a federally funded research and development center sponsored by the United States Army.

The Project Unique Identification Code (PUIC) for the project that produced this document is RAN126499.

(This page is intentionally blank.)

Contents

(This page is intentionally blank.)

Figures

Tables

(This page is intentionally blank.)

Summary

As the U.S. Army considers its future modernization and force structure options, it is useful for the service to compare and contrast its weapons systems, personnel management policies, operational concepts, and organization with those of other armies. While the U.S. Army is the fourth-largest and the best-equipped army in the world, there are areas where other militaries may have selected advantages. An understanding of the systems, concepts, and organizations of other armies will help the U.S. Army to focus its modernization effort and perhaps develop new operational concepts.

This project started in October 2012. The sponsor, the Deputy Chief of Staff, Force Development (G-8) in Headquarters, Department of the Army, asked RAND to examine a number of foreign armies in order to help inform the Army's modernization decisions, primarily in the Program Objective Memorandum (POM) period from 2014 to 2019. Despite the reality that no other army today has the depth and breadth of capabilities that the U.S. Army has, there was still a desire to take a selected look at what other armies were doing, primarily in terms of their material systems, in order to see if there were capability gaps present or emerging that the Army should consider addressing. Additionally, research could discover good ideas that other armies have that are not necessarily capability gaps in the U.S. Army, but are nevertheless worthy of consideration.

The organizing principle for the research was the Army's warfighting functions, formerly known as the battlefield operating systems. These functions include movement and maneuver (air and ground), intelligence, fires (indirect), sustainment, mission command, and protection. The comparison of the Army's systems with their foreign counterparts was performed within this framework. Due to the overall time and resource limitations of the research, within the warfighting functions, specific areas of focus were determined based on the needs of the sponsor. The specific areas that became the particular focus of the study included:

- Armored fighting vehicles and helicopters (movement and maneuver function).
- Multiple rocket launchers (MRLs) and towed and self-propelled cannons (fires function).
- Supply vehicles, watercraft, and engineer systems (sustainment function).
- Air and missile defenses—in particular, countering rockets (protection function).

Originally, there was the intent to examine the mission command (MC) function. It was determined that mission command was a highly technical area and would require specialized research and access to foreign systems; with the approval of the sponsor, the mission command comparison was moved to a separate study. A comparison of U.S. Army infantry squads with selected foreign counterparts replaced the MC effort.

The foreign armies used for the comparisons included those of U.S. allies (e.g., France, the United Kingdom, Germany, and Australia) as well as those of potential competitors or their potential suppliers (e.g., Russia and the People's Republic of China). In most cases, such as armored fighting vehicles, the research team initially cast a wide net to examine foreign systems. Once an understanding of the current state of the art was determined, a smaller collection of foreign systems was examined in much greater detail.

It is important to keep in mind that this research was mostly a systems and materiel comparison. The study focused on comparing U.S. Army material systems with their foreign counterparts, noting potential capability gaps as well as good ideas that other armies have that are worthy of consideration. Based on guidance from the sponsor, particular emphasis was placed on identifying potential or actual capability gaps that should be addressed in the POM period (from now to roughly 2020). Although the study was primarily systems oriented, there was an opportunity to compare foreign army organizations and operational concepts, and in selected areas to highlight how other armies compared with the U.S. Army. In order to keep the study unclassified, open sources were the primary source of information. Importantly, no modeling or simulation was conducted to support the research. The methodology employed was to identify and compare the open-source characteristics of the systems selected to include in the research while supplementing this information with classified sources where necessary.

Due to the very large range of topics that the study had to examine, a decision was made to conduct a broad overview of foreign systems at an appropriate level of detail that the sponsor felt appropriate and to focus on unclassified sources. Classified insights were provided to the sponsor on a selective basis, but it should be noted that this document is based entirely on unclassified, open sources.

Key insights from the research are summarized below. Details are provided in the chapters of the main body of the report.

The Operational Environment

Combat in Iraq and Afghanistan required well-protected vehicles. The threat of improvised explosive device (IED) attacks has affected vehicle design and protection philosophies. As a result, many systems—from armored fighting vehicles to logistics vehicles—have increased in weight. Heavier vehicles are not as easily transportable to theater and present a greater logistics burden in the operational area.

As the Army reorients its focus to the Asia-Pacific theater, the potential for urban warfare and IED attacks has not diminished, thus providing a strong argument for investing in a versatile combat fleet.

Ground Movement and Maneuver

The U.S. Army's armored fighting vehicles (AFVs) compare well with their foreign counterparts, particularly the M1A2 Abrams main battle tank, which is widely regarded as the world's best tank in terms of protection and antiarmor firepower. Because modern, second-generation forward-looking infrared (FLIR) sensors have become an increasingly common feature on many of the world's tanks and infantry fighting vehicles (IFVs), the U.S. Army should consider improving its existing FLIR systems or pursue other promising sensor technologies in order to maintain the advantage the Army's tanks and Bradleys have experienced in past direct-fire engagements. Additionally, the U.S. Army is an outlier among armies fielding main battle tanks given its lack of a dedicated high-explosive fragmentation (HE-Frag) round. The Advanced Multi-Purpose (AMP) round in development for Abrams will improve its ability to engage dismounted infantry and buildings and will eliminate one of the few areas where Abrams is at a disadvantage compared with a great many foreign main battle tanks.

The Army's Bradley infantry fighting vehicle is armed with a smaller 25-mm cannon compared with those of many foreign counterparts, which frequently mount 30-mm and 40-mm cannons. While the 25 mm has excellent armor penetration for its size, some of the larger cannons have superior antiarmor capabilities and substantially greater high-explosive payloads. Many foreign armies (for example, our British allies, who use the 30-mm armed Warrior IFV) use a higher-caliber gun than the U.S. Army does for the Bradley, which puts the Bradley at a relative disadvantage against some armored vehicles and well-constructed buildings. The Stryker armored personnel carrier is most commonly fielded with a .50 caliber machinegun, and is therefore even more lightly armed compared with foreign wheeled vehicles that are equipped with medium-caliber weapons.

Finally, while the U.S. Army does not currently field any air-droppable AFVs, several other foreign armies do. Accordingly, the U.S. Army may wish to consider adding such a vehicle to its suite of AFVs. Technology exists to make a highly mobile, well-armed, and moderately well-armored vehicle, albeit at the cost of heavy protection, but depending on the threat this could still be a good investment.

Movement and Maneuver (Helicopters)

The U.S. Army is the world's leader in term of the size and capability of its helicopter fleet. Foreign and U.S. attack and medium-lift helicopters alike have increased in sophistication. They are all built with composite materials and all-glass cockpits with liquid-crystal display (LCD) screens and multifunctional displays. The U.S. attack helicopter platforms have exhibited dominant target system capability due to their ability to receive and transmit unmanned aerial system (UAS) feeds and their use of third-generation FLIR. However, foreign attack and medium-lift helicopter platforms do have some niche advantages. For example, some foreign attack helicopters have avionics compatible with American standards, some have longer-ranged weapons, and foreign aircraft are less expensive than their U.S. counterparts. Some foreign medium-lift helicopters have higher payloads and longer maximum ranges because they use auxiliary fuel tanks.

The Army's heavy-lift helicopter, the CH-47F, has greater digital connectivity than its foreign counterparts, but it has a lower payload than the equivalent foreign systems. This is also the case when the CH-47F is compared with the Marine Corps heavy-lift helicopter, the CH-53K, which has a higher payload capacity and greater range, albeit at substantially greater cost, both to acquire and operate, compared with the CH-47F. Compared with foreign helicopters, both U.S. services have cutting-edge tactical digital connectivity that surpasses their foreign counterparts.

Indirect Fires

The U.S. Army's Multiple Launch Rocket System (MLRS) and the similar High Mobility Artillery Rocket System (HIMARS) lack the range of some of the heavy, foreign, large-caliber artillery rocket systems, particularly some that have been developed by China. Therefore, the rocket systems are falling behind the increasing range of similar Russian and Chinese rocket systems. The trend of foreign, heavy MRLs being able to fire well over 100 km has implications for the U.S. Army's fires system, including counterfire and target acquisition. Although the Guided MLRS (GMLRS) rocket has exceptional accuracy compared with any fielded foreign system, the suite of munitions available to MLRS and HIMARS is very limited compared with foreign rocket launchers. A large portion of the Army's current stock of rocket munitions will also have to be replaced when the 2019 limitations on submunitions take effect.

In terms of cannons, the U.S. Army's Paladin self-propelled howitzer has a digitized fire-control system, but lacks the high level of automation that exists in top-quality foreign self-propelled weapons such as the German PzH2000. The Army's towed M777 155-mm howitzer is a competitive weapon given its relatively low weight, which facilitates air transport. Finally, it should be noted that the Army's towed and self-propelled artillery are generally shorter-ranged than their modern foreign counterparts that use 45- and 52-caliber gun tubes.

Sustainment

The focus on counter-IED protection has led to progressively heavier logistics and engineering vehicles. These vehicles have also been called upon to perform other functions, such as surveillance and reconnaissance, which requires them to be deployable and have appropriate communications, weapons, and sensors, in addition to protection. Foreign armies are making increasing use of emerging technologies that improve 3-D imaging, engine performance, and lightweight armor and add remote acoustic sensors. However, increasing a vehicle's armor is still one of the predominant ways to improve survivability. In that regard, the trend toward better-protected logistics vehicles in the U.S. Army is mirrored in most of the foreign armies examined in this study.[1]

Within sustainment, the research also focused on watercraft. Most of the U.S. Army's watercraft (e.g., the Landing Craft Utility) fleet is based on mature, decades-old technology. Most watercraft (U.S. and foreign) have speeds of less than 20 knots. The one exception to that was hovercraft, which are used by several militaries examined in this study. While the U.S. Army does not operate hovercraft, the U.S. Navy's Landing Craft Air Cushion (LCAC) provides that capability within the U.S. armed forces.

Protection

The protection warfighting function was primarily assessed in terms of the threat posed by the long-range multiple rocket launchers that were mentioned in the section on indirect fires, as well as cruise missiles and unmanned aerial systems. That focus drove the research in the direction of gun, missile, and laser defensive systems.

A number of countries have or are fielding gun and missile defensive systems to protect against rockets, cruise missiles, and unmanned aircraft. Some of these systems are vehicle mounted, and therefore relatively mobile. By contrast, the Israeli Iron Dome counterrocket system is a generally fixed system intended to protect urban areas. A few countries, such as Germany, are working on ground-based laser defensive systems.

The conclusion reached in this functional area was that there was no clear standout in terms of defensive technology—at least not yet. Therefore, the U.S. Army should continue to fund research and development efforts to examine guns, missiles, and laser defensive systems until it becomes clear what the appropriate mix of systems will be.

Infantry Squads

As a result of more than a decade of combat experience in Iraq and Afghanistan, a number of armies have added considerable capabilities to their infantry squads. Radio communications for the individual soldier, improved body armor, better night vision equipment, new weapons, and man-portable jammers to protect against IEDs are becoming increasingly common in the world's infantry squads. The new kit, however, comes with a tactical price. All the armies that were consulted for this portion of the research (France, the United Kingdom, Australia, Canada, Germany, and Israel) indicated that the weight that infantrymen are now expected to carry is an increasingly important concern. Loads of 100 pounds or more are now entirely common for dismounted infantrymen. Several armies consulted for this study indicated that they are initiating programs to reduce the loads carried by infantrymen.

---

[1] In January 2019 the U.S. military will have to comply with new international norms on the use of submunitions. For example, explosive submunitions are prohibited unless they have a dud rate of less than 1 percent.

In terms of squad size, 8 to 12 was the range that was noted in the foreign comparisons. If the U.S. Marine Corps is included in the comparison, the Marine rifle squad is 13 personnel. Armies tend to size their infantry squads to match the size of their armored personnel carriers (APCs), but in the case of the German army there are two different sizes of infantry squads. Mechanized infantry (*panzergrenadiers*) that ride in the new Puma or older Marder IFVs are organized in seven-man squads, whereas light infantry squads that either walk or ride in unarmored trucks are in nine-man squads.

Unmanned Systems

The U.S. military has been investing in UASs and unmanned ground vehicles (UGVs) for the last few decades. The fighting in Iraq and Afghanistan dramatically accelerated that trend. While the United States and Israel have been at the forefront of tactical-level UASs, there is a relatively low cost of entry into the unmanned systems market, making it ripe for additional foreign competition. South Korea, for example, has developed an automated sentry that is employed along the Demilitarized Zone (DMZ) with North Korea. Iran and China are also rapidly moving into the UAS market.

Regarding UGVs, there has been fewer, but still important, strides in this field compared with UASs. Several nations, including Israel, have important niche advantages over the U.S. Army, such as armed UGVs that are used for surveillance and patrolling. Japan is currently the world leader in humanoid, bipedal robots that are coming increasingly close to having human-like mobility.

Acknowledgments

The authors would like to extend thanks to the sponsors at Headquarters, Department of the Army, G-8, for their support of the research. In particular, we wish to thank Lieutenant Colonel Douglas Waddingham, chief of operations for Director of Operations, Army G-8 Force Development (DOM), for his assistance in obtaining appropriate points of contact in the Army staff and his close, frequent interaction with the project team. A number of individuals provided important information that helped inform the research. At U.S. Army Training and Doctrine Command (TRADOC) headquarters at Fort Eustis, Jeffrey Hawkins, the foreign programs liaison officer, and Elrin Hundley of the Joint and Army Concepts Division were instrumental in establishing contact for the RAND research team with several of the allied liaison officers at Fort Eustis, who in turn provided exceptional assistance in obtaining information on their armies. Several representatives from the National Ground Intelligence Center (NGIC) in Charlottesville, Virginia, were very helpful, particularly in the areas of artillery systems and armored fighting vehicles, as were officials from the Maneuver Center of Excellence and the Fires Center of Excellence. British Army Lieutenant Colonel Paul Smith, assigned as an exchange officer to U.S. Army G-8 in the Pentagon, provided the research team with much information on the British Army, as well as additional British Army contacts who were very useful. Elizabeth Cole served as the project's administrative assistant and was instrumental in assembling this report. Finally, helpful reviews were provided by Jed Peters and Aaron Martin.

(This page is intentionally blank.)

Abbreviations

AAA	antiaircraft artillery
AAV7	Amphibious Assault Vehicle 7
ACRB	Advanced Chinook Rotor Blade
ADA	Air Defense Artillery
ADI	Australian Defence Industries
AFV	armored fighting vehicle
AGS	Armored Gun System
AHCAS	Advanced Helicopter Cockpit and Avionics System
AMAS	autonomous mobility appliqué system
AMP	Advanced Multi-Purpose
AMS	Avionics Management System
AMV	armored modular vehicle
APAS	Active Parallel Actuator System
APC	armored personnel carrier
APFSDS	armor-piercing fin-stabilized discarding sabot
APKWS	advanced precision kill weapons systems
APS	active protection system
APU	auxiliary power unit
ASE	Aircraft Survivability Equipment
ATACMS	Army Tactical Missile System
ATGM	antitank guided missile
ATK	Anti-Tank
BCT	brigade combat team

BUSK III	Bradley Urban Survival Kit III
C4I	Command, Control, Communications, Computers, and Intelligence
CAAS	Cockpit Avionics Architecture System
CAIC	Changhe Aircraft Industries Corporation
CAS	close air support
CCD-TV	Charge-Coupled Devices Television
CDAS	Cognitive Decision Aiding System
CDL	Common Data Link
CEP	circular error probable
CIWS	close-in weapons systems
CLP	combat logistics patrol
CMWS	Common Missile Warning System
COMFUT	Combatiente Futuro
COP	common operating picture
COTS	commercial off the shelf
C-RAM	counter rocket, artillery, and mortar
CROWS	Common Remotely Operated Weapon Station
CSAR	combat search and rescue
DaCAS	Digitally Aided Close Air Support
DAFCS	Digital Advanced Flight Control System
DAMO-FDL	logistics division, G-8
DARPA	Defense Advanced Research Projects Agency
DMZ	demilitarized zone
DoD	Department of Defense
DOM	Director of Operations, Army G-8 Force Development

DPICM	Dual-Purpose Improved Conventional Munitions
DSTL	Defence Science and Technology Laboratory
ECM	electronic countermeasure
EFV	Expeditionary Fighting Vehicle
EMP	electromagnetic pulse
ERA	explosive reactive armor
EW	electronic warfare
FADEC	full authority digital engine control
FIST	Future Integrated Soldier Technology
FLIR	forward-looking infrared
FMS	Foreign Military Sales
FMV	full-motion video
FPV	Future Protected Vehicle
FY	fiscal year
GCS	ground combat system
GCV	Ground Combat Vehicle
GFRP	glass-fiber reinforced plastic
GMLRS	Guided Multiple Launch Rocket System
GMLRS-AW	Guided Multiple Launch Rocket System Alternative Warhead
GPS	Global Positioning Service
GSAB	general support aviation battalion
HB	Hollow Base
HE	high explosive
HEAT	high-explosive antitank
HE-FRAG	high-explosive fragmentation

HFI	Hostile Fire Indicator
HIMARS	High Mobility Artillery Rocket System
HMLA	USMC Light Attack Helicopter
HMMWV	High Mobility Multipurpose Wheeled Vehicle
HMS/D	Helmet Mounted Sight and Display
FÉLIN	Fantassin à Équipement et Liaisons Intégrés
IAI	Israeli Aircraft Industries
IBCT	infantry brigade combat team
IED	improvised explosive device
IDF	Israeli Defense Force
IdZ-ES	Infanterist der Zukunft–Erweitertes System
IFV	infantry fighting vehicle
IICS	Integrated Infantry Combat System
INS	Inertial Navigation System
IRCCD	infrared charge coupled device
IRGC	Islamic Revolutionary Guard Corps
IRST	Infrared Search and Track
ISW	Indywidualny System Walki
IVHMS	integrated vehicle health management system
JAGM	Joint Air-to-Ground Missile
JATAS	Joint Allied Threat Awareness System
KMW	Krauss Maffei Wegman
kW	kilowatt
LCAC	Landing Craft Air Cushion
LCD	liquid-crystal display

LCT	Landing Craft Tank
LCU	Landing Craft Utility
LED	Light-Emitting Diode
LV	low velocity
MANET	Mobile Ad Hoc Network
MAWS	Missile Approach Warning System
MBAS	Mine Blast Attenuating Seats
MBT	main battle tank
MC	mission command
MDARS	Mobile Detection Assessment and Response System
MERS	Marine Expeditionary Rifle Squad
MGS	mobile gun system
MLRS	Multiple Launch Rocket System
MOD	Ministry of Defence
MPRS	Multi-Purpose Rifle System
MRAP	Mine Resistant Ambush Protected
MRL	multiple rocket launcher
MRSI	Multiple Round Simultaneous Impact
NBC	nuclear, biological, or chemical
NGIC	National Ground Intelligence Center
NLOS-C	Non-Line-of-Sight Cannon
NORMANS	Norwegian Modular Arctic Network Soldier
NREC	National Robotics Engineering Center
O&M	Operations and Maintenance
OSD	Office of the Secretary of Defense

PDA	Personal Digital Assistant
PGK	precision guidance kit
PGM	precision guided munition
PIM	Paladin Integrated Management
POM	Program Objective Memorandum
RAM	Rolling Airframe Missile
RC-IED	Remote Controlled—Improvised Explosive Device
RDT&E	research, development, testing, and evaluation
RFID	Radio Frequency Identification
ROW	rest of the world
RPG	rocket-propelled grenade
RWS	remote weapons system
S&T	science and technology
SBCT	Stryker Brigade Combat Team
SLWH	Singapore Lightweight Howitzer
SMSS	Squad Mission Support System
S-Nav	Soldier Navigation
SPAWAR	Space and Naval Warfare Systems Command
SRP	Software Reprogrammable Payload
SSC	Ship-to-Shore Connector
TARDEC	Tank Automotive Research, Development, and Engineering Center
THEL	Tactical High Energy Laser
TRADOC	U.S. Army Training and Doctrine Command
UAS	unmanned aerial system
UAV	unmanned aerial vehicle

UGV	unmanned ground vehicle
UHF	ultra-high frequency
UK	United Kingdom
USMC	United States Marine Corps
V-LAP	Velocity-Enhanced Long Range Artillery Projectile
VMF	Variable Message Format
WS-2	WeiShi-2

(This page is intentionally blank.)

1. Introduction

As the U.S. Army considers its future modernization and force structure options, it is useful for the service to compare and contrast its weapons systems, personnel management policies, operational concepts, and organization with those of other armies. While the U.S. Army is the fourth largest, and best-equipped, army in the world, there are areas where other militaries may have selected advantages. An understanding of the systems, concepts, and organizations of other armies will help the U.S. Army to focus its modernization effort and perhaps develop new operational concepts.

This project started in October 2012 at the beginning of fiscal year (FY) 2013. The sponsor, the Deputy Chief of Staff, Force Development, in the Pentagon, asked RAND to examine a number of foreign armies in order to help inform the Army's modernization decisions, primarily in the Project Objective Memorandum (POM) period from 2014 to 2019. Despite the recognition that no other army today has the depth and breadth of capabilities that the U.S. Army has, there was still a desire to take a systematic look at what other armies were doing, primarily in terms of their material systems, in order to see if there were capability gaps present or emerging that the U.S. Army should consider addressing. Additionally, it was recognized that the research could discover good ideas that other armies have that are not necessarily capability gaps in the U.S. Army, but are nevertheless worthy of consideration.

The organizing principle for the research was the Army's warfighting functions, formerly known as the battlefield operating systems. These functions include movement and maneuver (air and ground), intelligence, fires (indirect), sustainment, mission command (MC), and protection. The comparison of the Army's systems with their foreign counterparts was performed within this framework. Due to the overall time and resource limitations of the research, within the warfighting functions, specific areas of focus were determined based on the needs of the sponsor. The specific areas that became the particular focus of the study included:

- Armored fighting vehicles and helicopters (movement and maneuver function).
- Multiple rocket launchers (MRLs) and towed and self-propelled cannons (fires function).
- Supply vehicles, watercraft, and engineer systems (sustainment function).
- Air and missile defenses—in particular, countering rockets (protection function).

Since MC is such a broad and technical area, and would require specialized research and access to foreign systems, it was decided, with the approval of the sponsor, that the MC comparison would be performed in a separate study. A comparison of U.S. Army infantry squads with selected foreign counterparts replaced the MC effort.

A number of foreign armies were used for the comparisons. These included the armies of U.S. allies and friends (France, the United Kingdom, Germany, and Australia, for example) as well as potential competitors (e.g., Russia and the People's Republic of China). In most cases, such as armored fighting vehicles (AFVs), the research team initially cast a rather wide net to examine foreign systems. Once an understanding of the current state of the art was determined, a smaller collection of foreign systems was examined in much greater detail. Both open and classified sources were used. This report is an unclassified document. The classified data that the study utilized was retrieved from RAND. The sponsor was informed directly of classified insights and took note of those issues.

The specific tasks that RAND was asked to perform included:

1. Create a baseline framework and data to assess the current range of U.S. Army modernization programs. As noted above, the warfighting functions were used as the essence of this framework.
2. Determine the state of the field and the position of the core Army programs within their respective fields.
3. Assess the nature and importance of the relative position of U.S. Army capabilities.
4. Survey the rest of the world's ground militaries to identify new potential capabilities and program concepts.

Due to the potentially open-ended comparisons of U.S. Army systems with foreign counterparts in any of the warfighting functions, RAND asked the sponsor for guidance and input on areas of particular interest. For example, the sustainment function alone is a large portfolio, with essentially countless potential areas of comparison, such as equipment maintenance, food, fuel and ammunition supply, and transportation systems. Therefore, in this example, DAMO-FD, the Force Development division within G-8, was asked what areas were of particular interest to them that RAND could then focus on. In the case of Force Development division, they indicated that watercraft, mine, and improvised explosive device (IED) protection of supply vehicles and engineering equipment (particular in terms of the Army's portion of the Global Response Force) were areas of near-term concern. Using that guidance, RAND focused its efforts within the sustainment function.

It should also be noted that this research was mostly a broad overview comparison of systems and materiel. The study focused on comparing U.S. Army material systems with their foreign counterparts, noting potential capability gaps as well as capabilities or alternative approaches in other armies that are worthy of consideration. Based on guidance from the sponsor, particular emphasis was placed on identifying potential or actual capability gaps that should be addressed in the POM period. Although the study was primarily systems oriented, there was an opportunity to compare foreign army organizations and operational concepts, and in selected areas to highlight how other armies compared with the U.S. Army.

One of the limitations of this study was that there was little opportunity to examine possible joint alternatives to compensate for possible U.S. Army capability gaps in terms of how its systems directly compared with their foreign counterparts. For example, readers will see in the fires section that the Army's rocket launcher systems are falling considerably behind several foreign counterparts, especially in terms of range. Thus, in a strict system-to-system comparison, there could be an important capability gap for the Army. The study was not able to examine whether other U.S. joint capabilities (for example, U.S. airpower) would be able to compensate for a real or perceived Army capability gap.

Additionally, there were areas where there is no direct U.S. Army foreign counterpart to a foreign system. For example, the Russians and Chinese both have parachute-capable light armored vehicles in their airborne forces. The U.S. Army currently has no comparable system. That example was highlighted in the research as being both a possible capability gap for Army airborne forces and a good idea that the U.S. Army should consider adopting.

As of this writing, the U.S. Army is in the process of disengaging from more than a decade of irregular warfare in Afghanistan; the withdrawal from Iraq is already complete. In the future the Army must be prepared for missions that span the range of military operations, from low to high intensity, including hybrid warfare and operations in increasingly large, urban areas. Given that strategic-operational-tactical context, useful insights may be gained by examining foreign systems.

The following chapters provide details on the comparisons that were made, including recommendations for the Army to consider. While most of the information that is provided is systems oriented, doctrinal, training, and leader development insights are offered where it was deemed appropriate. The research conducted for this project did not include modeling or simulation to compare the capabilities of U.S. Army systems with their foreign counterparts at an engineering level of detail. The primary method used to develop comparisons were the on-the-record attributes of a system, such as range of weapons and the munitions they fired, weight and protection levels of vehicles, carrying capacity of vehicles either in terms of numbers of personnel or cargo, and the range and payload characteristics of helicopters.

(This page is intentionally blank.)

2. Ground Movement and Maneuver

Overall, the U.S. Army has significant advantages in their AFVs compared with their allies and adversaries. Indeed, most U.S. Army fighting vehicles are either at the cutting edge or far surpass their foreign counterparts. In the case of the Abrams tank, for example, any modifications we recommend are of a relatively minor nature. In this chapter, we discuss how the Army's main battle tanks (MBTs), tracked infantry fighting vehicles (IFVs), and wheeled armored personnel carriers (APCs) compare with selected foreign counterparts and provide implications for the future of these vehicles in U.S. Army service. When selecting comparable vehicles, we looked to what our allied counterparts such as the United Kingdom use, as well as other militaries who have fought alongside the U.S. Army in Iraq or Afghanistan. We also examined vehicles from countries like Israel, which face different threat environments and therefore require different modifications to their vehicles. Finally, we compared U.S. Army combat vehicles with those of Russia and China.

Key Trends in Armored Fighting Vehicles

The operational area of ground maneuver units is changing and becoming more complex for a number of reasons. First, units are now operating in much larger areas. Combat in Iraq and Afghanistan required troops to operate in a vast array of rural and urban settings while modifying their tactics and objectives accordingly. Second, combined arms organizations are now at the battalion level. Third, the proliferation of precision antiarmor weapons, which include direct and indirect fire, and IEDs have had a considerable impact on global AFV design. This has created demand for active protection systems (APSs), redesigned hulls, all-around protection, and onboard electronic countermeasures (ECMs). Older vehicles have required modifications to meet these standards, which comes at a cost to the Army.

Current combat operations in Afghanistan are focused on counterinsurgency and use irregular warfare methods and tactics to fight enemy forces. Even after the end of combat operations in Afghanistan, the United States and its allies anticipate that they will remain engaged in irregular warfare to some extent. Another important aspect of the conflicts in Iraq and Afghanistan is the focus on urban operations. U.S. Army representatives and others have maintained that in the post–Iraq and Afghanistan conflict environments, this same close, urban fighting is likely to remain the reality of the battlespace, and will continue to influence AFV design. With the continuing decline in emphasis on operations in Afghanistan, the U.S. Army's development of conventional warfighting capabilities is growing in importance.

One important consideration to note is that not all U.S. Army AFVs have a direct counterpart, so a point-to-point comparison with foreign systems is not always possible. In these situations, we examined the closest counterpart that had a comparable use for that country's military. For example, the Russians and Chinese both use parachute-capable light armor in their airborne units, and the Russians have some specialized vehicles to provide direct fire in urban areas. These vehicles will be discussed in greater depth later in this chapter.

Main Battle Tanks

The comparison of U.S. Army and foreign MBTs initially started with a broad examination of a large number of tanks currently in use around the world. Once the key attributes of modern tanks were understood, a smaller selection of vehicles was chosen for detailed comparison with the U.S. Army's M1A2 SEP v2 Abrams. Figure 2.1 shows the vehicles that were included in the MBT comparison.

Figure 2.1. Select MBTs for Comparison

M1A2 Abrams

Leopard 2A6

Merkava Mk IV

T-90A

SOURCES: (clockwise from top left) U.S. Army photo by Staff Sgt. John Couffer; photo by E._Heidtmann, CC BY 3.0; photo by Black Mammmba, CC BY 3.0; and photo by Vitaly V. Kuzmin, CC BY 3.0.

Current-generation MBTs vary in weight, from 45 to 70 tons. Protection has been given considerable priority over lethality or mobility by most countries, particularly Israel and the United States. All tank designs must balance the impact of heavy armor protection against mobility and transportability, and a number of different approaches have been taken to mitigate the necessary trade-offs between mobility, protection, and lethality. For example, a number of foreign tanks have been developed with an automatic loader replacing one of the crew members, which permits a smaller design for a given level of armor protection. The T-90A is an example of this type of tank, as is the French Leclerc. Western MBTs with four-person crews have other advantages, including improved situational awareness and the ability to maintain the vehicle. Countries such as the United States that have to deploy their MBTs to distant locations also have to consider the impact of heavy vehicles on the load capacity of the ships or planes that carry them overseas and the logistical requirements of 60–70-ton tanks. Countries such as Israel that engage in combat in their immediate vicinity need not be as concerned with transportability and can therefore install additional protective measures that increase the weight of the vehicles. Some general information about the specifications of these vehicles is provided in Table 2.1.

The Abrams is more than three decades old and the heaviest of the tanks we compared, but it is still the best tank in the world given its degree of armor protection and antiarmor capabilities. In Table 2.1, the introduction date of 1992 is the M1A2SEP variant of the Abrams; the original 105-mm-armed M-1 was introduced in service in the early 1980s.

Table 2.1. Main Battle Tanks—General Information

Vehicle	Country of Origin	Date of Introduction	Crew/ Dismounts	Status/Number in Service	Exported?	Combat Weight (short tons)
M1A2SEP V2 Abrams	USA	1992	4	Approx. 580 M1A2s and 580 M1A2 SEPs	Egypt, Kuwait, Australia	69.54
Merkava Mk 4	Israel	2003	4/8; ambulance version can carry 3 litter patients.	Approx. 400 in service; 300 more to be delivered	Merkava 4 not offered for export; Merkava 3 is	65
Leopard 2A6	Germany	2007	4	225 in German Army; more than 3,000 of all Leo 2 variants	Netherlands, Portugal, Canada, Spain, Greece	62.3
T-90	Russia	1995	3	743 in Russian Army	India (650), Algeria (305); Azerbaijan, Saudi Aabia, and Turkmenistan have placed orders	47.5

SOURCE: *Jane's Armor and Artillery.*

Mobility-Protection-Firepower Comparisons

Two Abrams variants are currently used by the U.S. Army: the M1A1 (which is in use in the Army National Guard, and a version is also used by the Marine Corps) and the M1A2.[2] Compared with the German Leopard 2A6 and the Israeli Merkava Mark IV, which have diesel engines, the Abrams has multiple fuel options for its turbine engine. While the use of a turbine means the Abrams has a higher rate of fuel consumption, especially when the tank is not moving, it also provides the tank with great acceleration capacity. Russia's T-90A tank also has multiple fuel options. The top speeds and cross-country performance of all the tanks that were included in this comparison were roughly similar, with the Israeli Merkava being the slowest of all four. The Merkava has the advantage of operating very close to its bases and logistics system. All the tanks included in this comparison are protected by modern, composite armor arrays.

The Abrams tank's use of depleted uranium armor gives it the best basic protection of any MBT. However, Israel's Merkava can be fielded equipped with the Trophy active defense system, which can detect, track, and destroy some antiarmor threats, including antitank guided missiles (ATGMs) and rockets launched far from the vehicle. Trophy also provides some degree of protection against top attack weapons for the Merkava.[3] The Leopard 2A6 can be equipped with appliqué top armor, and the T-90A's standard armor package contains modern explosive reactive armor, including some coverage of the front hull and turret roof. The Russian's Arena active defense system can also be fielded on T-90A and has been marketed for export but has not seen widespread fielding. Although both Trophy and Arena provide some ability to counter ATGMs missiles that employ top attack warheads (but approach horizontally), no fielded active defense systems were found that defend against more-vertical threats, such as artillery or air-deployed weapons.

The Abrams, Leopard 2, and Merkava all mount a similar 120-mm smoothbore cannon, though the Leopard 2A6 and later variants has a longer 55-caliber barrel. In terms of antiarmor capability, the shorter barrel on the Abrams is offset by the Army's use of superior ammunition, both in design and in composition (notably the use of depleted uranium in the M829 series of armor-piercing fin-stabilized discarding sabot [APFSDS] rounds). It is the use of depleted uranium ammunition that gives the Abrams the antiarmor advantage over the German Leopard 2A6.

The Army's M1A2s lack a high-explosive fragmentation (HE-Frag) round comparable to those fielded in many other countries. (The Marine Corps has adopted a German round, the DM-11, to provide an air-bursting high-explosive [HE] capability.) The Advanced Multi-Purpose (AMP) round in development would rectify this relatively small weakness in an otherwise superior platform. By comparison, the 125-mm main gun on the T-90A (similar to the main gun

---

[2] "General Dynamics Land Systems M1/M1A1/M1A2 Abrams MBT," *Jane's Armor and Artillery*, last updated April 3, 2012.

[3] "Merkava Mk 4 MBT," *Jane's Armor and Artillery*, last updated March 28, 2012.

on the T-72 and T-80 series tanks as well) falls short of the 120 mm in armor penetration, but has been able to fire a dedicated HE-Frag round since its development in 1962. Also of note, all of the tanks in this comparison were found to be equipped with second-generation forward-looking infrared (FLIR) systems. This notably includes the T-90A, which benefits from a French-developed FLIR system that was exported to Russia and is available for export. The latest version of the M1A2, called SEP V2, comes with the Common Remotely Operated Weapon Station (CROWS), which is equipped with sensors and thermal imaging that makes target acquisition and engagement from inside the vehicle possible while moving.[4]

Tracked Infantry Fighting Vehicles

IFVs have more variance in design philosophy compared with MBTs, which means countries can tailor them more specifically to meet their needs (see Figure 2.2). For example, today's IFVs range from 20 to more than 60 tons, with the trend moving toward heavier vehicles because of the emphasis on protection from IEDs and other urban combat threats. The U.S. Army's Bradley is in the middle of this weight range, and its weight has been growing in recent years. The gun armament of IFVs is mostly in the 30-mm–40-mm range, but gun caliber is increasing, as is the use of remote weapons systems (RWSs) in IFVs. ATGMs are a key armament variable; some armies include ATGMs on their IFVs, while others do not. Most countries build IFVs to carry a complete infantry squad, meaning these vehicles also serve an important transport function in addition to their weapons-carrying capabilities.[5]

We tracked IFVs from Israel, Germany, Sweden, and Russia for this comparison. Table 2.2 provides general information about each of these vehicles. As with the Abrams MBT, the Bradley IFV is the oldest of the vehicles in our examination but has undergone several modifications since its inception (the 1981 date listed in the table is when the first version of Bradley entered U.S. Army service). It also has the most vehicles in service compared with Israel's, Germany's, Sweden's, and Russia's IFVs, though it is not as widely exported as Sweden's CV90.

---

[4] "Common Remotely Operated Weapon Station (CROWS)," in Federation of American Scientists, *United States Army Weapons Systems 2013*, United States Army, pp. 78–79.

[5] The size of squads varies by army; the vehicles we examined carried squads of six to eight personnel (see Table 2.2).

Figure 2.2. Select IFVs for Comparison

U.S. M2A3 Bradley

German Puma

Israeli Namer

Russian BMP-3

Swedish CV90

SOURCES: (Clockwise, from top left) U.S. Army photo by Sergeant Quentin Johnson; photo by Sonaz, CC BY 3.0; photo by MathKnight and Zachi Evenor, CC BY 3.0; photo by Vovan, CC BY 3.0; and photo by BS, CC BY-SA 2.5.

Table 2.2. Tracked IFVs—General Information

Vehicle	Country of Origin	Date of Introduction	Crew/ Dismounts	Status/Number in Service	Exported?	Combat Weight (short tons)
M2A3	USA	1981	3 + 7	Production complete: 6,800+ built	M2A3 is U.S. only; 4,200 M2s to Saudi Arabia	39 with BUSK III
Namer	Israel	2008	3 + 8	In production: first batch is 250	None reported	68.3
Puma	Germany	2010	3 + 6	In production: first batch is 405	None reported	Up to 46.3
CV90	Sweden	1991	3 + 8	Production complete: 1,100+ built	Sweden, Denmark, Finland, Norway, Netherland, Switzerland	25-38.6 t
BMP-3	Russia	1990	3 + 7	In production: 1,600+ in service	Main users: Russia, UAE, Kuwait; several others in limited numbers	20.6 t

SOURCE: *Jane's Armour and Artillery.*

Mobility

The Bradley's weight with the Bradley Urban Survival Kit III (BUSK III), which provides features such as survivability seats and fire suppression systems, gives it a significantly lower power-to-weight ratio compared with the Puma, CV90, BMP-3, and even the Namer. As was the case with the Israeli Merkava MBT, Namer's weight is high, but since the vehicle is intended to fight close to its bases, weight is less of a concern; Namer is currently the heaviest IFV in use today.

To ensure that the Bradley would retain mobility, its engines and automatic transmission systems are to be upgraded to accommodate the extra weight BUSK added to the vehicle, but the programmed engineering change proposals to Bradley will only restore some of the vehicle's former performance.[6] The German Army, which faces a similar challenge with the weight of the Puma, equipped it with modular armor that can be removed to facilitate air transport. The variable protection levels of Puma mean that the vehicle's power-to-weight ratio will change, depending on the armor package that is installed. The BMP-3's use of light base armor gives it the highest power-to-weight ratio of the vehicles we examined. It has low ground pressure and is also amphibious. It is also relatively unique among IFVs in its incorporation of a rear final drive (most IFVs mount their engine in the front to preserve capacity in the rear), which improves its ride quality and weight balance.

Protection

BUSK III provides additional protection for Bradley based on experience gained in Iraq, including additional belly armor, changes to the fuel cell to reduce the likelihood of catastrophic kills, and additional reactive armor protecting the lower hull of the vehicle. The fact that Namer was developed from an existing MBT has resulted in it having MBT-like survivability, as it has the heaviest armor protection of any IFV. Namer has also been tested with the Iron Fist APS. To improve survivability in the German Puma, the vehicle is designed with the crew inside the hull compartment, with an unmanned RWS in the turret.

The Puma also uses a soft-kill APS, which employs sensors to confuse ATGMs by interfering with the missile's electronics.[7] Sweden's CV90, with its comparatively high power-to-weight ratio, maintains a low profile but good armor protection. Finally, Russia's BMP-3 employs only light-base armor with additional passive and reactive armor and an optional Arena APS. In this IFV, protection is secondary to mobility and cost-effectiveness, which provides insight into Russia's strategic decisions regarding what qualities are more important for an IFV—protecting the crew is less important compared with vehicle mobility and firepower.

---

[6] "BAE Systems M2 Infantry Fighting Vehicle/M3 Cavalry Fighting Vehicle," *Jane's Armour and Artillery*, last updated January 14, 2013.

[7] Ranjeet Singh, "Active Protection Systems," *South Asia Defence and Strategic Review*, July 26, 2011.

Firepower

The Bradley M2A3's primary armament is a stabilized 25-mm cannon with a coaxial machinegun and TOW-II ATGM. The vehicle is equipped with second-generation FLIR system for the gunner and an additional independent thermal viewer for the commander. The Namer has a .50-caliber heavy machine gun in a RWS; a 30-mm cannon version has been proposed but has not yet been fielded. On the Puma, the remote turret is armed with a 30-mm Mk44 weapons system. The German Army is interested in adding an ATGM (the Israeli Spike), but the vehicle is already equipped with second-generation FLIR and extensive provisions for the crew's situational awareness. The current version of the CV90 has a stabilized 40-mm cannon and a coaxial machinegun. The cannon can fire APFSDS rounds but lacks an ATGM.[8] The CV90 also has in its latest models an advanced fire control system with a second-generation FLIR and a customizable roof-mounted stabilized sight.[9] Finally, the Russian BMP-3 is armed with a 100-mm gun/missile launcher and a 30-mm auto cannon with 7.62-mm coaxial machinegun. This is the heaviest armament of any IFV, though the BMP-3's primary antitank capability comes from its ATGM. The vehicle can also be exported with an optional French-licensed second-generation thermal imager.

The research identified a number of foreign IFVs that use unmanned or remote weapon systems. This is an interesting concept that merits more examination for future fighting vehicles of all types.

Wheeled Armored Personnel Carriers and Infantry Fighting Vehicles

The U.S. Army's Stryker is primarily intended to fill the role of an APC for infantry as opposed to being a direct fire weapon (see Figure 2.3). Other armies use wheeled armored vehicles in reconnaissance and direct fire or light tank roles, in addition to using them as APCs. These vehicles fall in a range from 12 to 30 tons, with Stryker in the middle of that range.

---

[8] "Combat Vehicle 90 (CV90) (Stridsfordon 90) Infantry Fighting Vehicle," *Jane's Land Warfare Platforms: Armoured Fighting Vehicles*, updated November 21, 2011.

[9] Combat Vehicle 90 (CV90) (Stridsfordon 90) Infantry Fighting Vehicle," 2011.

Figure 2.3. Select Wheeled Armored Vehicles for Comparison

U.S. M1126 Stryker ICV

French VBCI

Finnish Patria AMV

Ukraine BTR-4

SOURCES: (Clockwise, from top left) U.S. Navy photo by Mass Communication Specialist 1st Class Daniel N. Woods; photo by Daniel Steger, CC BY-SA 2.5; "MIL_Finlândia-Army_Demo Day 2005 Rovajärvellä" shared by MATEUS_27:24&25 via Flickr, CC BY-SA 2.0; and publicity photo from Kharkiv Morozov Machine Building Design Bureau.

Stryker is based on the LAV-III, which has been exported in different models to other countries, and Finland's Patria armored modular vehicle (AMV) has also been exported to numerous European countries and has been used in Afghanistan. In Table 2.3, the 2002 date for Stryker is when the first versions started to enter U.S. Army service.

Table 2.3. Wheeled Fighting Vehicles—General Information

Vehicle	Country of Origin	Date of Introduction	Crew/ Dismounts	Status/Number in Service	Exported?	Combat Weight (short tons)
Stryker	USA	2002	Infantry Carrier Vehicle: 2 + 9	In production: 4,000+, all variants	Widely available as LAV-III/Piranha III	17.2; up to 27.5 with new drive line and suspension
AMV	Finland	2003	APC Variant: 1 + 10	In production: 1,400+, all variants	UAE, Poland, South Africa, Croatia, Slovenia, Sweden, Spain	26.45 (amphibious at 24.25)
VBCI	France	2008	3 + 9	In production: 630	None to date	20 (stretch potential to 32)
BTR-4	Ukraine	2009	3 + 8	In production: 10 in Ukraine	Macedonia; 420 ordered by Iraq	19.3 to 29.76

SOURCE: *Jane's Armour and Artillery.*

Mobility

The U.S. Army's Stryker has a 450-horsepower engine, providing it with good road mobility for a vehicle of its weight. The 2011 version of the vehicle was also fitted with a 55,000-pound suspension system and a smart power-management system that directs the Stryker's power needs, sending power to the appropriate parts of the vehicle where and when it is needed to maximize efficiency.[10] One of the most exported AFVs in Europe, the Patria AMV also has a good power-to-weight ratio and is transportable by the new European A400M four-engine turbo-prop cargo plane. By contrast, France's VBCI is heavier and not as easy to transport by air, though its skid steering enables sharp turns, which is a useful feature in urban environments. The last vehicle we examined, Ukraine's BTR-4, is fully amphibious.

Protection

In 2011 General Dynamics delivered the first Strykers with a double-V hull, putting their survivability on par with the Mine Resistant Ambush Protected (MRAP) vehicles. Slat armor is set to be replaced by explosive reactive armor (ERA) to better protect the vehicle from RPG-7–type weapons and IEDs. Stryker also comes equipped with Mine Blast Attenuating Seats (MBAS). The Finnish AMV has a modular design that allows users to customize their protection levels based on their needs. Seats are also hung from the roof and the sides of the vehicle to

---

[10] Lance M. Bacon, "Stryker Gets Another Round of Upgrades," *The Army Times*, last updated August 6, 2012.

ensure maximum crew survivability should the vehicle hit a mine.[11] This same feature is also present in the BTR-4, which also comes in an up-armored version. Finally, the VBCI is the only wheeled IFV we examined that comes with top-attack protection kits.

Firepower

The Stryker has a 40-mm MK-19 automatic grenade launcher or a .50-caliber M2 HB (Hollow Base) machine gun in its CROWS RWS. The AMV's RWS comes with various calibers of cannons available and one .50-caliber heavy machine gun; a version of the AMV (the Rosomak) was developed and fielded by Poland with a 30-mm cannon in a manned turret. The VBCI has a turret-mounted 25-mm M811 dual-feed cannon, smoke grenade launchers, and a coaxial-mounted 7.62-mm machine gun. Finally, the BTR-4 is armed with a 30-mm cannon and one 7.62 coaxial machine gun in a remote turret, along with six 81-mm grenades.[12] The remote turret adds nearly two tons of weight to the vehicle.[13] It should be noted that when compared with its foreign wheeled fighting vehicle counterparts, the infantry carrier version of Stryker is lightly armed since it lacks an automatic cannon.

Airborne Light-Armored Fighting Vehicles

The U.S. Army's ability to air-drop light armor in support of airborne forces formerly consisted of a battalion of M551 Sheridan light tanks in the 82nd Airborne Division; this vehicle was retired in 1996, and its replacement, the M-8 Armored Gun System (AGS), was canceled in the same year. Air transportability via tactical air transports such as the C-130 was also a feature of the Future Combat Systems manned ground vehicles. With the cancelation of the Future Combat Systems, the Army lacks a light-armored vehicle able to provide direct fire support that is capable of deployment via airdrop. At the time of this writing, this capability is being examined by the Maneuver Center of Excellence under the name Mobile Protected Firepower in response to a statement of need issued by XVIII Airborne Corps earlier this year.

In light of the Army's consideration of the readoption of an air-droppable light-armored fighting vehicle, this project examined a number of foreign air-droppable fighting vehicles. These include most notably a family of airborne light-armored vehicles that has gone through several generations of development from Russia, and also a number of other systems that have been adopted with this mission in mind, including Chinese, French, and German systems.

Russia's airborne forces possess a variety of tracked light-armored vehicles based on the BMD, which is currently in its fourth generation (BMD-4M being the latest version). The BMD-

---

[11] "Patria Land Systems Armoured Modular Vehicle," *Jane's Armour and Artillery*, last updated November 30, 2011.

[12] "GROM Universal Fighting Module," *Kharkiv Morozov Machine Building's Design Bureau*, undated.

[13] "Kharkov Morozov Design Bureau BTR-4 Armoured Personnel Carrier," *Jane's Land Warfare Platforms: Armoured Fighting Vehicles,* last updated January 26, 2012.

4M, which can be airdropped, is a fully amphibious IFV, has excellent cross-country mobility, and possesses identical armament to BMP-3. Figure 2.4 provides additional information regarding the BMD-4M's specifications.

Figure 2.4. Russia's BMD-4M—Technical Specifications

Dimensions
Length: 6.1m (238 in.)
Width: 3.1m (121 in.)
Height: 2.4m (94 in.)
Weight: 13,600 kg (29,956 lbs)

Crew size
         3, with 5-6 dismounts

Mobility
Power: 500 HP engine (same as BMP-3) Power-weight ratio: 36.7 hp/ton
Road speed: 44 mph
Swim speed: 6 mph

Armament
100mm rifled main gun
- Fires HE to 7,000m (34 rounds)
- Able to fire AT-10 laser-guided ATGM to 6,000m
30mm coaxial autocannon
- Max range with HE: 4,000m
- Max range with AP: 2,500m
- 7.62mm coaxial machinegun

SOURCE: Photo by Vitaly V. Kuzmin, CC BY-SA 3.0.

As of December 2012, the commander of the Russian airborne troops announced plans to reequip the four divisions and one separate brigade of the airborne branch with the BMD-4M.[14] Other vehicles based on the BMD chassis include a mortar carrier (2S9 Nona) equipped with a turreted, breech-loading 120-mm mortar, and an armored personnel carrier variant (BTR-MD Rakushka) that can carry up to 13 infantry under light armor. A light tank, the 2S25 Sprut-SD, has also been developed that is armed with a 125-mm smoothbore main gun derived from the T-90A's cannon.

China has fielded its own version of the BMD, a vehicle called ZBD-03, which is armed with a 30-mm cannon and ATGM launcher. As is the case in Russia, China's airborne forces are a strategic rapid reaction force—the Chinese 15th Airborne Corps, consisting of three divisions, is part of the People's Liberation Army Air Force and answers directly to China's senior leadership. Further development of Chinese airborne forces may see more-sophisticated airborne fighting vehicles fielded.

The key element in the design of AFVs is the constraint that air-droppability places on their dimensions and weight, and therefore the protection levels of these vehicles. None of the vehicles mentioned here is well protected against direct fire systems beyond heavy machine

---

[14] "Russia to Commission BMD-4M Airborne Vehicles in 2013," *RiaNovosti,* December 27, 2012.

guns, and several are only protected against small arms. At best, they have comparable protection against IEDs and mines to the M113. Their light weights give them excellent tactical mobility, and several of them are very heavily armed.

Implications for the U.S. Army

Overall, the U.S. Army's AFVs compare favorably with their foreign counterparts, though a few niche areas were identified where foreign capabilities have approached parity with or exceeded the capabilities of U.S. systems. Building off the Army's incremental modernization processes and focus on refitting existing technology to improve its fleet,[15] we offer three low- to medium-level priorities for consideration in the POM time frame.

This research identified the proliferation of modern, second-generation FLIR systems throughout the current generation of foreign MBTs and IFVs, including those available on the market to potential adversaries. Improvements to vehicle sensors will be necessary to regain the battlefield advantage that the Army enjoyed in 1991 and 2003 due to its early adoption of first- and second-generation FLIR systems. The Army should invest research and development funds and expand on existing technology to preserve and extend the current tactical advantage these vehicles have in direct fire capability in all weather and visibility. Improvements in sensors and direct fire targeting have the potential to substantially affect multiple classes of combat vehicles, including the Army's MBTs, the Bradley and its eventual replacement, and other systems employing stabilized direct fire weapons, such as the Stryker mobile gun system (MGS). Research and development options that the Army could consider include (1) how to degrade the increasingly capable FLIRs that are appearing on foreign systems, and (2) if there other direct fire sensor technologies that could supplement or replace FLIRs in order to maintain the Army's direct fire advantage.

Another area of possible improvement for the Army's IFVs and wheeled fighting vehicles is to increase the size of its main armament. The trend in foreign tracked and wheeled IFVs is toward automatic cannons in the 30–40-mm class. Another vehicle that was initially examined in this research was the British Warrior. The Warrior is currently armed with a 30-mm cannon. As part of the program to extend the Warrior's service life, the British Army is going to refit the vehicle with a 40-mm cannon. Bradley's 25-mm Bushmaster is a good weapon, but against some armored vehicles and well-constructed buildings, the weapon is at a disadvantage compared with the heavier weapons found on most modern foreign vehicles. Strykers fight as part of a combined arms team at the tactical level, where they would often have support from other systems, including the Abrams MBTs, which have a much more powerful direct fire armament than any Stryker variant. That said, there is no denying that when compared with its foreign counterparts,

---

[15] *A New Equipping Strategy: Modernizing the U.S. Army of 2020*, national security report, torchbearer issue, Arlington, Va.: Institute of Land Warfare and Association of the Untied States Army, June 2012, p. 20.

the main version of Stryker (the infantry carrier) is very lightly armed. Therefore, it should be in the Army's interest to examine the armament of foreign wheeled fighting vehicles and conduct experiments to determine if a heavier armament is warranted. Should the Army elect to retain the 25 mm on the Bradley until a new IFV is fielded, it may be possible to improve the gun's ammunition as an interim measure, rather than moving to a larger-caliber gun, as many foreign armies are doing.

The U.S. Army is an outlier among major countries fielding MBTs due to its lack of a dedicated HE-FRAG round. While U.S. 120-mm antitank rounds are without peer anywhere in the world, the Army lags both friendly and potential adversary countries by decades in fielding an air-bursting HE round. The 120-mm AMP round in development will result in the Army having this capability. Should it be required sooner, or if development of this round is delayed or canceled, ammunition is available from multiple countries that is compatible with the Abrams and, in one case, has already been fielded by the U.S. Marine Corps on its tanks.

In light of the interest by XVIII Airborne Corps in an air-droppable AFV, this research identified several countries that have fielded light airborne AFVs capable of providing direct fire support to parachute forces. The size and weight requirements to ensure air-droppability force significant trade-offs in vehicle design, but it is demonstrably possible to develop a vehicle that can be air-dropped, is highly mobile, and is armed with up MBT armament, albeit with light-armor protection. This is another area where the U.S. Army should consider following the lead of foreign armies.

3. Indirect Fires

This chapter provides comparisons of the main U.S. Army indirect fire systems with their foreign counterparts. The analysis focused on towed and self-propelled medium howitzers and rocket artillery systems. In each case, the munitions that the foreign systems use were included in the assessment.

Self-Propelled Howitzers

The M109A6 Paladin (Figure 3.1) is the U.S. Army's sole self-propelled howitzer, and it equips all of the Army's armored brigade combat teams. Compared with the field as a whole, the M109A6 Paladin is a solid performer in a few key respects but lacks the more powerful gun and automation of the current generation of modern howitzer systems, which results in it lacking in range and burst rate of fire relative to many foreign systems.

Figure 3.1. M109A6 Paladin

SOURCE: Photo by the U.S. Army.

The M109A6 Paladin ([Paladin Integrated Management [PIM] version pictured in Figure 3.1) entered service in 1992 and is based on the M109 chassis, which was originally fielded in 1963.[16] The Army has unsuccessfully attempted to replace Paladin twice in the past decade, most recently with the Crusader advanced self-propelled howitzer and the Non-Line-of-Sight

---

[16] "BAE Systems US Combat Systems M109A6 155 mm Paladin Self-Propelled Howitzer," *Jane's Armour and Artillery*, updated February 7, 2012.

Cannon (NLOS-C) Future Combat Systems vehicle. The age and relatively low mobility of Paladin are seen as its most significant weaknesses, and as such current Army plans call for a major upgrade, entitled PIM. PIM mainly consists of a new chassis for Paladin, built by BAE Systems and featuring substantial commonality with the Bradley family of vehicles. This should help mitigate sustainment and mobility issues but will not address limitations in Paladin's range and rate of fire. In both Operation Desert Storm (in 1991) and Operation Iraqi Freedom (in 2003), Paladin-equipped artillery units had some difficulty keeping pace with armored units armed with Abrams and Bradley tanks and IFVs.

Figure 3.2. Selected Foreign Self-Propelled Howitzer Systems

K9 Thunder

PzH 2000

PLZ-05

CAESAR

SOURCES: (Clockwise, from top left) photo by Defense Citizen Network, CC BY-SA 2.0 Korea; photo by the Dutch Ministry of Defence, CC BY-SA 1.0; photo by Max Smith, and photo by Daniel Steger, CC BY-SA 1.0 Generic.

The K9 Thunder is the Republic of Korea's new self-propelled howitzer, and it is produced by Samsung-Techwin (see Figure 3.2).[17] It is being fielded in South Korea and

---

[17] "Samsung Techwin 155 mm/52-Calibre K9 Thunder Self-Propelled Artillery System," *Jane's Armour and Artillery*, updated February 7, 2012.

Turkey (as T-155 Firtina, with some locally produced components). Like PzH 2000, the K9 Thunder has a longer, more powerful gun and a greater degree of automation compared with Paladin. It is included here as a representative self-propelled artillery system due to the numbers being produced (some 1,500); it is similar to other systems being fielded that are roughly a generation more advanced than Paladin.

PzH 2000 is the German Army's main self-propelled howitzer, and it has also been fielded by Italy, the Netherlands, and Greece.[18] PzH 2000 was fielded in 1998 and represents the state of the art in modern self-propelled guns. It is manufactured by Krauss Maffei Wegman (KMW) and features a more powerful gun; substantial automation, which permits a smaller crew size and higher rate of fire; and improved survivability and internal ammunition capacity. As a result, it is also close to twice the weight of Paladin, but still has a superior power-to-weight ratio, owing to a much more powerful engine. Importantly, PzH 2000 has a much greater range and significantly higher rate of fire compared with Paladin.

The PLZ-05 is a Chinese-built modern howitzer that is included here as a system that could potentially be marketed to various future adversary nations.[19] The previous Chinese self-propelled gun, the PLZ-45, has already been marketed to several countries, and it shares a number of characteristics with the Russian 2S19 MSTA self-propelled howitzer. Of particular note, this howitzer is not based on the old Soviet 152-mm but instead mounts a NATO-style 155-mm gun.

CAESAR is a new wheeled howitzer system fielded by the French Army in 2008 that has already seen action in Afghanistan and Mali.[20] It has also been exported to Saudi Arabia and Thailand and is being actively marketed by Nexter Systems. It is included here as an alternative self-propelled system that manages to combine automation and a longer, more powerful gun than that of Paladin, with a lightweight chassis (weighing under 20 tons at combat weight).

Lethality

Table 3.1 provides additional details regarding the lethality of each system discussed in this chapter. Paladin mounts a 39-caliber 155-mm main gun,[21] but the other systems included for comparison all are equipped with longer and more powerful 52-caliber guns. These guns have

---

[18] "Krassu-Maffei Wegmann Pazerhaubitze 2000 (PzH 2000)," *Jane's Armour and Artillery*, updated February 7, 2012.

[19] "NORINCO 155 mm Self-Propelled Gun Howitzer PLZ-05," *Jane's Armour and Artillery*, updated February 7, 2012.

[20] "Nexter Systems CAESAR 155 mm Self-Propelled Gun," *Jane's Armour and Artillery*, updated February 7, 2012

[21] Where artillery dimensions are concerned, *caliber* is the multiple of the length of a gun barrel compared with its diameter; hence a 39-caliber gun is 39 times as long as it is wide.

longer barrels and larger chambers, which permits more-powerful charges to be used. The 39-caliber 155-mm gun on Paladin (and also on the M198 and M777 towed howitzers) has an 18-liter chamber, which can accommodate less propellant than the (generally) 23-liter chambers of 45- and 52-caliber guns that are in wide circulation elsewhere in the world. Medium self-propelled howitzers that outrange Paladin while firing standard ammunition are increasingly common. While Paladin can make up some of the range gap using Excalibur, this requires the use of an expensive round that is fielded only with a unitary warhead; the disadvantage of Paladin's shorter range while employing special ammunition types, such as smoke or illumination rounds, remains.

Table 3.1. Self-Propelled Howitzer Lethality Comparisons

Name	Main Armament	Range		Maximum Rate of Fire	Ammo Storage
		Conventional	Extended		
155-mm 39 caliber	155-mm 39 caliber	22.6 km	40 km (Excalibur)	4/min	39
K9 Thunder	155-mm 52 caliber	30 km	55–60 km	8/min 3 in 15 s	48
PLZ-05	155-mm 52 caliber	30 km	50+ km	8/min	30
PzH 2000	155-mm 52 caliber	30 km	55–60 km	10/min	60
CAESAR	155-mm 52 caliber	30 km	55–60 km	6/min; 3 in 18 s	18

SOURCE: *Jane's Armor and Artillery*.

Another feature of modern artillery systems is a high degree of automation. Paladin has a digitized fire control system and a hydraulic rammer, but some systems, such as PzH 2000, have much higher levels of automation. This permits a high burst rate of fire—up to ten rounds a minute with some systems. Due to heating of the barrel, the sustained rate of fire on all modern howitzers falls to approximately two rounds per minute over time, but for brief periods, automated systems provide a significant advantage. This extends to the ability to conduct Multiple Round Simultaneous Impact (MRSI) missions, where an individual howitzer with automated loading and laying mechanisms is capable of firing multiple rounds with trajectories and timing that enable the rounds to reach the same target at the same time. PzH 2000 has demonstrated the ability to fire a five-round MRSI against a target 17 km away. Other systems have at least some capability to carry out MRSI; K9 is claimed to be able to fire three-round MRSI fire missions, and one South African system, G6-52, is asserted to be capable of six-round MRSI at 25 km.

A quick comparison of the ability of a platoon of four Paladins and four PzH 2000s to deliver fires over a three-minute period shows the limitation of the U.S. system compared with the leader among the world's self-propelled howitzers. While a Paladin platoon could deliver

48 shells in an intense three-minute fire mission, the German platoon could deliver 120 shells—and could do so at distances up to 50 percent greater than Paladin's maximum range.

Towed Howitzers

The Stryker Brigade Combat Team (SBCT) and some fires brigades are equipped with towed 155-mm howitzers; the current version, the M777, is replacing the M198 in these units (see Figure 3.3). With the planned reorganization of the infantry brigade combat teams (IBCTs), a battery of 155-mm howitzers will also be part of the brigade's artillery battalion. Compared with the field as a whole, the M777 is among the most lightweight and modern towed medium howitzers, though it lacks the range or automation of other heavier, towed artillery systems.

Figure 3.3. U.S. Army M777 and M198 Towed 155-mm Howitzers

M777   M198

SOURCE: Photos by the U.S. Army.

The M777 howitzer is a towed 155-mm howitzer with a weight of about 9,000 pounds.[22] The latest version, M777A2, is capable of firing M982 Excalibur ammunition. It is in use in the Army and Marine Corps as well as in Australia and Canada, with a Foreign Military Sales (FMS) order by Saudi Arabia made in 2011. The lighter weight of this system enables transport by a wider range of vehicles, including medium-lift helicopters like the UH-60. Its incorporation into the IBCT will also ensure that for the first time all brigade combat teams (BCTs) have an organic precision fires capability at up to a range of 40 km.

The M198 howitzer is an older 155-mm howitzer system in use by the U.S. Army. Its larger size and weight of about 15,800 pounds somewhat limits the prime movers that can be

---

[22] "BAE Systems, Global Combat Systems 155mm Lightweight Howitzer (M777)," *Jane's Armour and Artillery*, updated March 12, 2012.

used to tow it, as well as restricting it to air transport by CH-47. In addition to its U.S. use, it is fielded in ten other countries.[23]

Figure 3.4. Selected Foreign Towed 155-mm Howitzers

GHN-45

SLWH Pegasus

155/52 APU SBT

SOURCES: (Clockwise, from top-left) photo by Sturmvogel 66, CC BY-SA 3.0; photo by MINDEF Singapore; and photo by Outisnn, CC BY-SA 3.0.

The GHN-45 is an improved version of the GC 45 155-mm howitzer that was developed by South Africa (see Figure 3.4).[24] This version was developed by Austria, and about 600 have been made. It has been fielded by Iran and Thailand. It is an example of a towed howitzer that can be fitted with an auxiliary power unit (APU) that offers some limited ability to self-propel the gun and to enable automated reloading. Its longer barrel also provides it a greater maximum range than that of the M777, but it is also a significantly heavier weapon, at 27,300 pounds.

The 155/52 APU SBT is an example of one of the relatively few 52-caliber towed 155-mm howitzers. It is the largest towed howitzer examined here, and weighs almost 30,000 pounds. Its APU provides power for raising and lowering the wheels, opening the trails, and operation of the gun. As a result, General Dynamics European Land Systems claims that the gun can be brought into action with only two minutes of preparation, and it is capable of a maximum range and rate of fire comparable to that of modern self-propelled howitzers.[25]

---

[23] "155 mm howitzer M198," *Jane's Armour and Artillery*, updated March 12, 2012.

[24] "NORICUM GH N-45 155 mm Gun-Howitzer," *Jane's Armour and Artillery*, updated March 1, 2012.

[25] *Advanced Artillery System: SIAC 155/52*, brochure, Madrid: General Dynamics European Land Systems, January 2012.

The SLWH (Singapore Lightweight Howitzer) Pegasus is the closest analog to the M777; it is about 3,000 pounds heavier than the M777 and has an identical 39-caliber 155-mm cannon, but is equipped with an APU that powers an automatic loading system. It has been fielded by the Singapore Armed Forces since 2005, but has not been sold elsewhere. Table 3.2 compares the lethality of towed howitzers.

Table 3.2. Towed Howitzer Lethality Comparisons

Name	Main Armament	Range		Maximum Rate of Fire	Ammo Storage
		Conventional	Extended		
M777/M198	155-mm 39 caliber	22.6 km	40 km (Excalibur)	4/min	N/A
SLWH Pegasus	155-mm 39 caliber	22.6 km	30 km	4/min, 3 in 24s	N/A
GHN-45	155-mm 45 caliber	24.7 km	39.6 km	7/min	N/A
155/52 APU SBT	155-mm 52 caliber	30 km	55–60 km	10 in first min; 3 in 11s	N/A

SOURCE: *Jane's Armor and Artillery*.

While the more powerful 45- and 52-caliber guns have been adopted with increasing frequency in self-propelled howitzers, these are less common among towed guns. The 155/52 APU SBT has been fielded in relatively limited numbers (around 80) by Spain and Colombia. Guns similar to the 155/52 APU SBT have been fielded in Singapore (FH2000), Turkey (Panter), and Finland (GH 52) in smaller numbers. Some of the more powerful Russian 152-mm guns, such as the 2A36, have similar characteristics but appear to lack a comparable suite of projectiles that would enable them to engage targets beyond 35 km.

The use of an APU on towed howitzers appears to be an increasingly common feature. A powered, automated loading system adds weight but permits a higher rate of fire and, potentially, a decreased crew requirement. It can also offer limited mobility to a deployed gun.

Cannon Ammunition

This section is intended to provide more context for a discussion of trends in modern 155-mm artillery ammunition. Given the vast number of different types of artillery rounds, we tailored this section to focus on efforts to extend effective range and precision, since those are key characteristics of artillery ammunition.

A number of extended-range but unguided munitions have been developed or are in development that take advantage of the higher chamber pressures and longer barrels of 52-caliber guns. One example is the LU 211 round being developed by Nexter Systems for use with France's CAESAR wheeled self-propelled howitzer. The LU 211 HB round can reach a maximum range of 30 km, and it can be converted to a base bleed projectile in the field, which grants a maximum range of about 40 km.[26]

Denel has also been developing a Velocity-Enhanced Long Range Artillery Projectile (V-LAP), which can serve as an example of the shell in a way to improve aerodynamic performance. Figure 3.5 gives Denel's reported performance for the V-LAP compared with base bleed and boattail projectiles. V-LAPs are rocket assisted shells. Base bleed ammunition shells have a cavity in the rear of the round that is filled with a combustible material that fills in the vacuum that forms behind a shell in flight, thus decreasing the drag on the projectile. Boattail ammunition is shaped toward the rear of the shell in a way to improve aerodynamic performance.

Figure 3.5. Effects of Ammunition Types on Maximum Range, in Kilometers

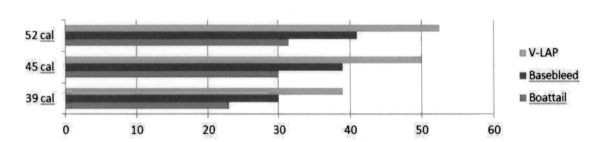

SOURCE: Denel, "155mm V-LAP Round," *Jane's Ammunition Handbook*, updated January 25, 2013.

This round has been fired to 56 km from a PzH 2000, and in one test has achieved 75 km when fired from a Denel G6-52. When compared with conventional HE rounds, the V-LAP sacrifices a significant amount of explosive filler (4.5 kg, compared with 10.8 kg in M795).

The primary precision artillery round currently employed by the U.S. Army is the M982 Excalibur 155-mm round. This round employs GPS/Inertial Navigation System (INS) guidance to achieve 10-m circular error probable (CEP) accuracy while also extending the maximum range of Paladin and 155-mm towed guns through the use of an advanced design incorporating glide fins. It has a maximum range of up to 40 km when fired from a 39-caliber gun, and up to 60 km when fired from a 52-caliber gun.

In addition to purpose-built precision munitions like Excalibur, the United States has also pursued a precision guidance kit (PGK) that can improve the precision of an existing "dumb" artillery shell. Three efforts to develop PGKs for current artillery projectiles were found; one offered by Anti-Tank (ATK) that has been supported by the U.S. Army, one by Nexter that is programmed for adoption by the French Army by 2015, and one called TopGun that is under

---

[26] Christopher F. Foss, "France Seeks Out Improved Artillery Projectiles," *Jane's International Defense Review*, July 9, 2012b.

development by Israeli Aerospace Industries that claims better than 10-m CEP at any range.[27] All of these PGKs are special fuses that offer the ability to guide conventional artillery rounds that are currently in service; they are less expensive than specialized guided rounds. Unlike Excalibur, however, PGK fuses slightly reduce a shell's maximum range.

Rocket Artillery

The U.S. Army has fielded a rocket artillery system that encompasses two launcher vehicles. In the present day these are excellent weapon systems; they have high accuracy and competitive ranges using the current rockets, and those systems that outrange them are much heavier and so far not available in large numbers. However, this is an area where the U.S. Army may risk falling behind if current trends persist, and where some potential weaknesses, particularly a limited suite of ammunition types, may mean that foreign rocket systems have capabilities that Army systems do not.

Figure 3.6. U.S. Army Rocket Artillery Systems

M270 MLRS   M142 HIMARS

SOURCES: (left to right) U.S. Army photo and U.S. Army photo by Staff Sgt. Rafael Andrade.

M142 High Mobility Artillery Rocket System (HIMARS) is the newer of two artillery rocket systems in use in the U.S. Army (see Figure 3.6). It fires a single pod of six 227-mm rockets or one Army Tactical Missile System (ATACMS) missile.[28] With the fielding of the GPS-guided M30 and M31 Guided Multiple Launch Rocket System (GMLRS) rockets, it has

---

[27] See "TopGun," *IAI.com*, undated

[28] "Lockheed Martin Missiles and Fire Control 227 mm Multiple Launch Rocket System (MLRS)," *Jane's Armour and Artillery*, updated July 25, 2013.

a fairly long range and high accuracy. It entered service in 2002 and more than 400 have been produced.

The M270 Multiple Launch Rocket System (MLRS) is a tracked, armored rocket launcher that carries two pods of rockets or missiles.[29] It was first fielded in 1983 and more than 1,200 have been produced. It is in service in at least fourteen other countries, not counting the United States.

Figure 3.7. Selected Foreign Multiple Rocket Launcher Systems

BM-21 Grand   BM-30 Smerch

Fajr-5

Lynx

SOURCES: (Clockwise, from top left) photo by Robert Wray, CC BY-SA 3.0; photo by Michael, CC BY 3.0 Unported; photo by M-ATF, CC BY-SA 3.0; publicity photo from Israel Military Industries.

---

[29] "Lockheed Martin Missiles and Fire Control 227 mm Multiple Launch Rocket System (MLRS)," *Jane's Armour and Artillery*, updated July 25, 2013.

The BM-21 Grad is a 122-mm multiple rocket launcher (MRL) that is, counting variants and locally produced copies, the most commonly used MRL platform in the world by a significant margin (see Figure 3.7).[30] At least 5,000 have been produced since its introduction in 1963, and it remains in service in about 60 countries. Although lighter and shorter ranged than most of the general support rocket systems considered here, it is included both due to its ubiquity and in order to note that improved 122-mm rockets have the potential to significantly upgrade the capability of this widely available system.

The BM-30 Smerch is a mature, long-range, 300-mm artillery rocket system.[31] It has been fielded by Russia as well as at least eight other countries, and more than 300 have been made. A Chinese 300-mm rocket system called PHL-03 also appears to be a copy of this system. It is the most capable former Soviet-era rocket system. Both BM-21 and BM-30 employ a tube-based rocket system that can only be reloaded one rocket at a time, rather than in sealed pods as with MLRS, HIMARS, and other modern systems. Currently, BM-30 slightly outranges the U.S. Army's MLRS and HIMARS firing GMLRS, and given present Russian development efforts, that situation is likely to worsen in the near future.

The Fajr-5 is a 333-mm indigenously produced Iranian rocket launcher system, firing four rockets from tubes, or a single double-stage long-ranged variant.[32] It was fielded in the 1990s. Fajr-5 rockets have been employed by Hezbollah in limited numbers. Its accuracy is poor by U.S. standards, but its range is comparable to MLRS and HIMARS.

Lynx is an advanced multicaliber rocket system developed by Israel.[33] It is capable of firing a range of munitions, including 122-mm and 220-mm Russian artillery rockets, the Israeli 160-mm LAR-160 and 306-mm EXTRA rockets, and the Delilah-GL cruise missile system. Some of its rockets have also been designed to be compatible with Israeli MLRS launchers.

WeiShi-2 (WS-2) is one of a number of large and very long-range Chinese artillery rockets.[34] It is a 400-mm rocket system that was fielded in 2007, but detailed information on the numbers produced is not available. These and similar rocket systems are being produced and marketed by multiple Chinese arms manufacturers. Presently, WS-2 outranges the rockets fired by MLRS and HIMARS by considerable margins, and that situation is likely to worsen in the coming years.

---

[30] "SPLAV 122mm BM-21 Multiple Rocket Launcher Family," *Jane's Armour and Artillery*, updated July 23, 2013.

[31] "SPLAV 300 mm BM 9A52 (12-Round) Smerch Multiple Rocket System," *Jane's Armour and Artillery*, updated July 23, 2013.

[32] "333 mm Fadjr-5 Iranian Rocket," *Jane's Armour and Artillery*, updated February 5, 2013.

[33] "Lynx Autonomous Multi-Purpose Rocket System," Israeli Military Industries, undated.

[34] "SCAIC 400 mm WS-2 Multiple Rocket Weapon System," *Jane's Armour and Artillery*, updated July 22, 2013.

One of the key discriminators between the MLRS launcher and the other rocket systems discussed here is that it is the only tracked vehicle system. In terms of lethality characteristics, however, the principal focus has been on the munitions: rocket payload size and content, range, and accuracy are the key factors, and any of the rockets discussed here can be fired from armored or unarmored and tracked or wheeled vehicles.

The U.S.-developed and fielded GMLRS rocket currently leads the field in accuracy and has a fairly long effective range compared with most foreign systems. It is very accurate and its main limitations are the fact that, given Department of Defense (DoD) guidance relating to cluster munitions, it is presently only available with a unitary warhead, and it is an expensive munition. Future developments to GMLRS include an alternative warhead (GMLRS-AW) and a new rocket with an enhanced range of around 250 km. Given the trends in Chinese and Russian MRLs, a longer-range MLRS rocket for the U.S. system is an important addition.

Figure 3.8 shows the ranges of major rocket artillery and counterbattery radar systems (the U.S. Q-37 and EQ-36 shown on the chart) in roughly the 1980s, the present day, and ten years from now, with current U.S. capabilities included for comparison. It demonstrates three key trends:

1. U.S. artillery rocket ranges have improved significantly over time, but without a continued emphasis on further increases in range, GMLRS will begin to be eclipsed by the latest Russian and, especially, Chinese rocket systems.
2. The entrance of the Chinese and their greater emphasis on much heavier, longer-range rockets that begin to bridge the gap between rocket artillery and short-range ballistic missiles could have a significant effect over time in extending the trend toward longer-range strike systems.
3. Limitations in current and projected counterbattery radar systems could mean that detection, warning, and counterbattery targeting may need to be handled through new capabilities and procedures in Army Air Defense Artillery (ADA) or joint radars, including airborne systems.

Figure 3.8. Example Artillery Rocket Ranges: Past, Present, and Future

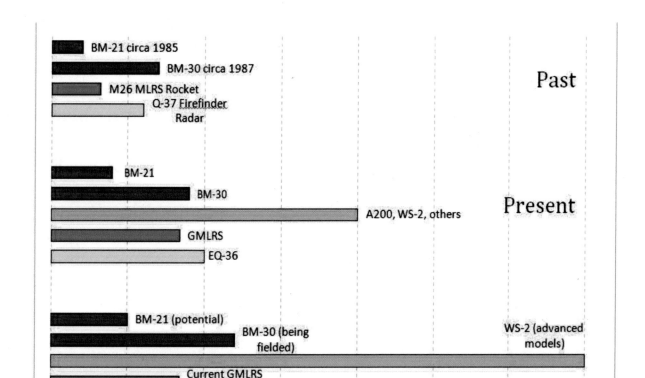

Range (km)

Range is only one aspect of artillery rocket capability, though it is a critical one. As is shown in Table 3.3, GMLRS has the advantage in accuracy, but some high-end foreign systems have an advantage in range, and a substantially greater selection of potential warhead types.

Table 3.3 focuses on some of the higher-end potential threat systems fielded by Russia and China, but the trends that are suggested here—increasing range, accuracy, and variety of warhead types—have been growing over time. The most common systems, such as BM-21, are typically still fielded with older rockets, which are inexpensive and available in large quantities, but versions that leverage GPS guidance and advanced propellants in order to extend their range and improve accuracy are already available.

Finally, the variety of warhead types available for foreign rockets is substantial compared with that of the MLRS family of rockets in its present form. In foreign use, these heavy rockets are typically under the control of corps or higher-level headquarters; the emphasis has remained on conventional missions, and the ability to contribute usefully in an environment with limited rules of engagement is a lower priority for the Russian and Chinese militaries.

Table 3.3. Key Artillery Rocket Capabilities

	M270A1 MLRS M142 HIMARS	9A52 Smerch (BM-30)	WeiShi 2 (WS-2)
Range	15–84 km	Current generation of rockets has a range of 25–90 km, with 120 km rockets in development	Currently fielded with 200-km range; latest variants with ranges of approximately 350 km
Accuracy	GPS/INS guidance of 5–10-m CEP	Course correction available: 0.23 percent of range claimed (~207 m at 90 km)	GPS/INS guidance available 200-m CEP claimed
Typical warhead size	90 kg	95–100 kg	200 kg
Warhead types available	Unitary High Explosive DPICM (limited use) Germans have a steerable mine warhead	4 types of submunition warheads 2 types of precision submunitions antitank mines Fuel-air explosive parachute–retarded HE-Frag Hardened HE (earth penetrating)	DPICM-type submunitions Comprehensive effect submunitions Fuel Air Explosive HE-Frag Incendiary Unmanned aerial systems

SOURCE: *Jane's Armor and Artillery.*
NOTE: DPICM = Dual-Purpose Improved Conventional Munitions.

Implications for the Army's Indirect-Fire-Procurement Decisionmaking

It is perhaps important to note that in the area of indirect fires, various armies have differing doctrines and philosophies regarding the role and use of cannons and rocket launchers. In the case of the U.S. Army, artillery cannons are primarily used as a weapon to provide direct support to infantry and armor, with the MLRS and HIMARS rocket launchers used mostly for counterfire against enemy artillery. Both missions are tactical in nature. Some of the very long-range foreign rocket launchers, such as the Chinese WS-2, have operational-level potential. It is no coincidence that the range of the WS-2 allows it to fire across the Taiwan Strait. In the U.S. military, a long-range strike mission of that type would be conducted by aircraft.

While there are limits to the extent to which recommendations can be made based on trends in artillery systems worldwide, our research has uncovered some information that can inform decisions about a number of ongoing programs in the fires portfolio.

Cannons

Regarding self-propelled howitzers: While PIM will improve mobility and ease the sustainment burden of the M109A6 Paladin, it will not appreciably address the issue of Paladin's range or rate of fire. Future upgrades to the gun and level of automation of this system will be necessary to bring U.S. artillery capabilities in line with the best foreign self-propelled guns currently fielded. The continuing excellence of U.S. ammunition development with the Excalibur is also worth noting, as no comparable artillery round was found to exist elsewhere.

Wheeled self-propelled guns have recently gained in popularity and a system comparable to the French CAESAR might be a feasible path toward providing more mobile indirect fire support to the Army's SBCTs.

The fielding of the M777 has given the Army an excellent towed medium howitzers. The addition of an auxiliary power unit to improve mobility, responsiveness, and rate of fire could be explored, with the caveat that such additions would inevitably add weight and complexity to the system.

Rockets

The Army's MLRS and HIMARS units provide the Unite States with a good fire support capability. Those foreign systems with superior range and payload are much heavier systems fielded in more limited numbers and by fewer countries. GMLRS accuracy is the best among the world's current artillery rockets.

However, the trend is clearly toward longer-range rocket systems, and improvements to rocket munitions can proliferate quickly among countries that already own the launcher vehicles. Efforts to extend the range of GMLRS and to ensure a sufficient service life for the MLRS- and HIMARS-fired ATACMS would help keep Army fires systems abreast of this trend. Indeed, the trend toward MRLs with ranges of 100 to possibly over 300 km is an important one for the U.S. Army to watch in terms of its own similar systems (MLRS and HIMARS and their associated ammunition), as well as its artillery-locating capabilities. This could require new Army and joint solutions, such as longer-range ground-based radars and an aerial artillery-locating capability on either manned or unmanned aircraft.

The GMLRS Alternative Warhead that replaces DPICM will help restore some of the gap in the variety of warhead types available to U.S. rocket artillery units, but a more complete suite of modern warhead payloads would require further investment and attention beyond GMLRS-AW.

(This page is intentionally blank.)

4. Helicopters

This chapter analyzes a selection of attack, medium-lift, and heavy-lift helicopters in order to support the Army with its equipment-modernization investment decisions. We primarily looked at open source materials to examine a variety of military helicopters used by a number of foreign militaries. The platforms in Table 4.1 were included in the analysis.

Table 4.1. Helicopters Types and Platforms Included

Attack Helicopters	Medium-Lift Helicopter	Heavy-Lift Helicopters
AH-64E Apache Guardian AH-1Z Viper	UH-60M Blackhawk Eurocopter EC725 Caracal Mi-171A2	CH-47F Chinook CH-53K Super Stallion Mi-26T2
Eurocopter EC665 Tiger Mi-28N Havoc Z-10		

Although this chapter does not include an analysis of specific scout/reconnaissance helicopters, the team did conduct a literature survey for this class of helicopter to see what was happening with existing inventories and whether new scout helicopter platforms were being developed around the world. The literature search of this class of aircraft allows some conclusions to be drawn about how the U.S. Army's reconnaissance helicopter programs compare with foreign counterparts. Those insights will be provided at the end of this chapter.

As was the case with fires systems, it should be noted that different militaries have varying doctrines and philosophies regarding the use of helicopters. The U.S. Army has had a strong aviation branch since the Vietnam War in the 1960s, when thousands of helicopters were used. Today the U.S. Army has helicopter units in aviation brigades that are habitually associated with specific divisions (and prior to today's organizational construct, Army helicopters were organic to divisions). In most other armies, helicopters are less numerous, and they can even be controlled and operated by the country's air force. Even within the U.S. military there are different tactical employment concepts for attack helicopters. For example, the Marine Corps views its Cobra attack helicopters as a close air support (CAS) asset controlled and allocated in a manner similar to fixed-wing CAS aircraft. The Army, on the other hand, doctrinally views attack helicopter units as a maneuver unit similar to traditional ground maneuver forces such as tanks and infantry.

Selection Criteria

Various foreign helicopter platforms were considered,[35] but specific platforms were chosen based on the priority of the system to the user nation (for example, the Mi-28 was chosen over the Ka-50 and Mi-24 because the Russian military has decided that the Mi-28 is its attack helicopter of the future). We also focused on state-of-the-art systems (as well as older systems undergoing extensive modernization upgrades) and similar mission aircraft (with search and rescue and antisubmarine helicopters excluded from the analysis since they have no U.S. Army counterparts); geographic representation; and the desire to include aircraft of friends and potential adversaries. Additionally, the analysis included several helicopters used by the Marine Corps.

Attack Helicopters

AH-64E General Platform Information

The AH-64E (Figure 4.1) is a four-blade, twin-engine, tandem, all-glass-cockpit attack helicopter manufactured by Boeing. It is known as the "Apache Guardian" (formerly the Apache Block III); it is the latest version of the Apache helicopter used by the U.S. Army, and it is a truly state-of-the-art platform. The first AH-64E was delivered to the Army in November 2011 and full-rate production was achieved in October 2012. As of January 2013, the Army had received 28 aircraft—enough to field its first full unit—but it plans on procuring a total of 690 through AH-64D model remanufacturing.

Figure 4.1. AH-64E Apache

SOURCE: Photo by the U.S. Army.

---

[35] The following helicopters were considered: Attack Platforms Ka-50 Hokum, Mi-24 Hind, A-129 Mangusta, and AH-2 Rooivalk, and Transport Platforms Super Puma/Cougar, Lynx, CH-46, and Super Frelon.

The AH-64E is armed with a variety of weapons, including a 30-mm chin-mounted cannon, 2.75" unguided rockets, advanced precision kill weapons systems (APKWS) laser-guided rockets, and Hellfire laser-guided missiles. The AH-64E also incorporates several new technologies designed to enhance the aircraft's maneuverability, survivability, and lethality, such as an improved drive system, composite rotor blades, new engines enhanced by digital control units, the ability to control unmanned aerial system (UAS) navigation and payloads through the high-bandwidth UAS Tactical Common Data Link (CDL) Assembly, digital interoperability with many fixed-wing aircraft through a Link 16 data link, the Cognitive Decision Aiding System (CDAS) to reduce pilot workload, the dual-mode seeker Joint Air-to-Ground Missile (JAGM) for improved weapon-to-target match and extended range, the Joint Tactical Radio System for improved connectivity across waveforms, and fully integrated aircraft survivability equipment (ASE).[36]

Marine Corps AH-1Z

The AH-1Z (Figure 4.2) is also a four-blade, twin-engine, tandem, all-glass-cockpit attack helicopter manufactured by Bell. Known as the "Viper" (or the "Zulu Cobra"), the AH-1Z is the latest version of the AH-1 helicopter used by the U.S. Marine Corps, and like the Apache, it is a state-of-the-art platform. The first AH-1Z was delivered to the Marine Corps in January 2007 and full-rate production was achieved in December 2010. As of April 2013, the Marine Corps had received 30 Vipers, but it plans on procuring a total of 226 through new builds and AH-1W model remanufacturing.

---

[36] See *2014 Army Equipment Modernization Plan*, Headquarters, Department of the Army, May 13, 2013; Amy Butler, "U.S. Army Prepares for Full-Rate AH-64E Production," *AviationWeek.com*, October 26, 2012; John Keller, "Boeing Moves Apache Block III Attack Helicopter Program Forward with $187 Million Army Contract," *MilitaryAerospace.com*, March 18, 2012a; Dave Majumdar, "US Army Fields First AH-64E Unit, but More Improvements to Come," *FlightGlobal.com*, January 9, 2013; Kris Osborn, "Technology Gives Apache Block III More Lift, Capability, Landing Ability," *Army.mil*, February 26, 2010; "Boeing, US Army Mark Delivery of 1st AH-64D Apache Block III Combat Helicopter," news release, *Boeing.com*, November 2, 2011; Stephen Trimble, "From Albania to Afghanistan, US Army Integrates Lessons into Latest Apache," *FlightGlobal.com*, November 3, 2011; Sydney J. Freedberg Jr., "Army Plays Shell Game with Unfinished Apache Helicopters: Put the Transmission in, and Pull It out Again," *Breakingdefense.com*, April 26, 2013.

Figure 4.2. AH-1Z Viper

SOURCE: U.S. Navy photo by Mass Communication Specialist 3rd Class Dominique Pineiro.

The AH-1Z is armed with a 20-mm chin-mounted cannon, 2.75" unguided rockets, APKWS laser-guided rockets, Hellfire laser-guided missiles, and AIM-9 Sidewinder air-to-air missiles. Like the AH-64E, the AH-1Z incorporates several new technologies designed to enhance the Viper's maneuverability, survivability, and lethality. However, unlike the AH-64E, the Viper is part of a larger United States Marine Corps (USMC) modernization initiative called the H-1 Upgrade Program, which in addition to 226 Vipers, will deliver 160 state-of-the-art UH-1Y "Venom" utility helicopters—both of which are employed in USMC Light Attack Helicopter (HMLA) squadrons—that have 84 percent parts commonality with the Viper.

These upgraded technologies include composite rotor blades that more than double the payload capacity of the AH-1W, a composite rotor hub that has 75 percent fewer moving parts than four-blade articulated rotor hubs, the Topowl Helmet Mounted Sight and Display (HMS/D) system with fourth-generation day, night, and night vision goggle visor-projection technology; an advanced targeting system with third-generation FLIR that can identify targets in excess of the aircraft's organic maximum weapons ranges; an integrated avionics system with an open architecture approach for 50 percent growth capacity; and a number of digital interoperability improvements, such as high-bandwidth CDL to send and receive full-motion video (FMV) and associated metadata, a digitally aided close air support (DaCAS) system for the digital exchange of CAS mission information, the next generation of Blue Force Tracker, and the Joint Allied Threat Awareness System (JATAS), which makes onboard ASE systems interoperable with the joint force.[37]

---

[37] See Bell Helicopter, *H-1 Program: AH-1Z and UH-1Y*, No. 1, 2012–2013; Amos, *Commandant of the Marine Corps: 2013 Report to the House Armed Services Committee on the Posture of the United States Marine Corps*, House Armed Services Committee, April 16, 2013; "Bell Helicopter AH-1Z Earns Navy Approval for Full Rate

European EC665

The Eurocopter EC665 Tiger (Figure 4.3) is four-blade, twin-engine, tandem, all-glass-cockpit attack helicopter manufactured by Eurocopter. The Tiger first became operational in the French and German militaries in 2005. As of December 2012, each country operated nearly 80 Tigers and each plans on procuring a total of 120 aircraft. In 2007, the Tiger became operational in the Australian and Spanish militaries, and each country currently operates approximately 20 Tiger variants. France, Germany, Australia, and Spain each use the Tiger in a slightly different manner based on emphasis given to air-to-air, air-to-ground, general support, or armed reconnaissance missions, and each military has procured slightly different subsystems tailored for their needs.

Figure 4.3. Eurocopter Tiger

SOURCE: Photo by besopha, CC BY-SA 2.0.

However, 77 percent of all Tiger variants' structural weight are components of carbon, aramid, or fiberglass, which allows an overall weight reduction up to 30 percent, corrosion-free structures, damage-tolerant behavior, and easy battlefield repair.

All Tiger variants are armed with a 30-mm chin-mounted cannon, but different variants use a mix of 70- and 68-mm unguided rockets; HOT, Trigat, Spike, Pars 3, and Hellfire antitank missiles; and Mistral and Stinger air-to-air missiles. Other modern technologies include the Thales Topowl HMS/D system (French versions); the BAE Systems day and night HMS/D (German version); the Australian Defence Industries (ADI) HMS/D system (Australia version); enhanced electromagnetic pulse (EMP) protection due to copper and bronze bonding throughout the fuselage; a mast or roof sight with an infrared charge coupled device (IRCCD) camera and laser rangefinder; a nose-mounted FLIR; an EADS Defence Electronics electronic

Production," *Shephard News*, December 10, 2010; "AH-1W/AH-1Z Super Cobra Attack Helicopter, United States of America," *Army-Technology.com*, undated; *H-1 Program: AH-1Z and UH-1Y*, Naval Air Systems Command, PMA-276 pamphlet, 2013.

warfare suite that includes a radar warning receiver; a missile launch detector; and a chaff/flare dispenser.[38]

Russian Mi-28N

The Mi-28N "Havoc" (Figure 4.4) is a five-blade, twin-engine, tandem all-glass-cockpit attack helicopter manufactured by Mil Helicopters. It is a dedicated attack helicopter, and unlike many armed Russian helicopters, it has no intended secondary transport capability. The first production aircraft was delivered to the Russian military in 2008, and the following year, the Mi-28N was delivered to its first operational unit. As of October 2012, there were 64 Mi-28N helicopters in service, and that number is expected to climb to 200 by 2015, which is the point at which it will completely replace the aging Mi-24 Hind.

Figure 4.4. Mi-28N Havoc

SOURCE: "Mil Mi-28N Havoc" shared via Flickr by Alan Wilson, CC BY-SA 2.0 Generic.

The Mi-28N is the world's heaviest attack helicopter due, in part, to its heavily armed cockpits and a windshield that is able to withstand hits from 12.7-mm to 14.5-mm caliber rounds. The Mi-28N is armed with a 30-mm chin-mounted cannon that features side-mounted ammunition boxes to prevent jamming. The 30-mm gun is also the same gun installed on the BMP-3 IFV, BMD-2 AFV, BMD-3 AFV, and BTR-90 APC. The Mi-28N can also carry 80-mm and 122-mm unguided rockets, AT-9 SACLOS and AT-12 laser-guided antitank missiles (with high-explosive antitank [HEAT] and thermobaric warheads), and SA-16 or AA-11 air-to-air missiles.

---

[38] See "Unlike Other Attack Helicopters in Its Class, the Eurocopter Tiger Sits the Pilot in the Front Cockpit and the Weapons Officer in the Rear Cockpit," *MilitaryFactory.com*, June 8, 2011; "Tiger Multi-Role Combat Helicopter, Germany," *Army-Technology.com*, undated; "Characteristics," Tiger, *Eurocopter.com*, undated; "Eurocopter 665 Tiger/Tigre," *Jane's Defence Equipment and Technology*, May 13, 2013; "Chapter Four: Europe," *The Military Balance*, Vol. 113, No. 1, 2013.

Other modernization technologies include composite rotor blades that can withstand hits up to 30 mm, an HMS/D system that can hand off targets to the navigator's surveillance and fire control system, an integrated electronic combat system that includes a microwave radar antenna mounted above the rotor head (similar to the AH-64D, however this system is not currently being delivered to operational units), a FLIR, integrated laser designator and tracker, low-light level television, and digitized flight, systems, and target information on liquid crystal displays (LCDs). The Mi-28N is also built with enhanced survivability features, such as engines with suppressed exhaust that enable a 2.5-times reduction in thermal signature from the Mi-24 predecessor and an emergency jettisoning system that blows off stub wings and canopy doors to allow for pilot bailout (pilots wear parachutes). The Mi-28N also has an ASE suite that includes a laser warning receiver, flare dispensing system, radar jammer, and radar and missile warning receivers; however, the aircraft is currently being delivered to flying units without the installed ASE equipment.[39]

Chinese Z-10

The Chinese Z-10 (Figure 4.5) is a five-blade, twin-engine, tandem, all-glass-cockpit attack helicopter manufactured by Changhe Aircraft Industries Corporation (CAIC). The Z-10 is the first indigenous Chinese attack helicopter; however, the Chinese had considerable assistance from the Russians to develop the aircraft. The Chinese also had assistance from by Pratt & Whitney Canada, which provided the aircraft's engines before the company was fined US$75 million by the U.S. government for a violation of export rules. As of January 2013, the Chinese had fielded 48 Z-10 helicopters, which are organized into four different squadrons, each of which has 12 helicopters.

The Z-10 is armed with a 23-mm cannon, 57- or 90-mm unguided rockets, HJ-9 antitank guided missiles (comparable to the TOW-2A), HJ-10 antitank missiles (comparable to the AGM-114 Hellfire), and TY-90 air-to-air missiles. Z-10 helicopters, however, are seldom observed with weapons, and as of late 2012, they have never been seen in flight with a full weapons load. The Z-10 has a millimeter-wave fire-control radar (like the Apache Longbow), the Infrared Search and Track (IRST) system, a low-light level television camera, and a laser designator. Both pilots have HMS/D with integrated night vision optics, although export variants are not offered with this capability. The Z-10 is outfitted with a fully integrated ASE

---

[39] See "Mil Mi-26 and Mi-27," *Jane's Defence Equipment and Technology*, June 24, 2013; "Russian Military to Purchase 10-15 Mi-28N Helicopters per Year," *RiaNovosti*, January 22, 2008; "Russian Air Force Receives First Mi-28 Night Hunter Helicopter," *RiaNovosti*, June 5, 2006; "Mi-28A/N Havoc Attack Helicopter, Russian Federation," *Army-Technology.com*, undated; "Mil Mi-28 Havoc Attack Helicopter," *Military-Today.com*, undated; "The Mil Mi-28 Havoc Has Since Become the Standard Attack Helicopter for the Russian Air Force and Army," *MilitaryFactory.com*, last updated February 26, 2014; Air Force Tactics, Techniques, and Procedures (AFTTP) 3-1, "Rotary-Wing Aircraft, Employment, and Tactics," in *Threat Guide: Threat Reference Guide and Countertactics*, pp. 8–56, December 3, 2012; "Chapter Five: Russia and Eurasia," *The Military Balance*, Vol. 113, No. 1, 2013.

system and an electronic warfare (EW) system that can degrade incoming tracking signals through a jamming pod. Although the Z-10 is assessed to be a formidable force within the next decade—particularly due to the inclusion of government-mandated critical technologies—the Chinese will likely experience several years of growing pains with this aircraft, as evidenced by problems associated with the Z-10's engines.[40]

Figure 4.5. Z-10

SOURCE: Photo by Shimin Gu, GNU Free Documentation License, Version 1.2.

General Characteristics of Attack Helicopters

Table 4.2 presents the general characteristics of attack helicopters.

Table 4.2. General Characteristics of Attack Helicopters

Helicopter	Operational Introduction	Number in Service	Export Location	Max Weight/ Max Payload	Max Speed (max/cruise)	Max Range
AH-64E Apache Guardian	2012	690 (planned)	Taiwan and South Korea; possibly Qatar, UAE, and Saudi Arabia	23,000 lbs	158/143 knots	260 NM
AH-1Z Viper	2010	226 (planned)	South Korea	18,500 lbs 5,764 lbs	155/134 knots	370 NM
Eurocopter EC665 Tiger	2003	France: 120 (planned) Germany: 120 (planned)	Australia, UK, Netherlands, and Spain	14,553 lbs 3,968 lbs	150 124 knots	430 NM
Mi-28N Havoc	2009	200 (planned)	Iraq and Kenya	25,350 lbs 5,180 lbs	172/145 knots	234 NM
Z-10	2009	48 (currently)	Unknown	15,432 lbs	161/135 knots	430 NM

---

[40] "Z-10 Attack Helicopter, China," *Army-Technology.com*, undated.

Attack Helicopter Platform Trends and Insights

The section presents trends and insights regarding attack helicopter platforms. There is increased sophistication in U.S. *and* foreign platforms: All modern attack helicopters are being built with:

- Composite materials
- Fly-by-wire (FWB) electronic controls
- All-glass cockpits with large-screen, liquid-crystal, multifunctional displays
- Helmet-mounted targeting displays
- Hellfire class missile systems
- Integrated and advanced ASE suites
- Enhanced survivability.

Regarding U.S. dominance in targeting system capability: U.S. attack helicopters have a significant lead in this area due to:

- Ability to receive and retransmit FMV UAS feeds
- Sensor-to-weapon match: third-generation FLIRs with greater target identification ranges.

Regarding U.S. dominance in weapons system capability: U.S. attack helicopters have significant lead in the area due to:

- Lower-yield, lower-cost PGM options (APKWS II)
- Development of dual-mode seeker for PGM (JAGM).

Foreign platform shortfalls can be compared with U.S. attack helicopters:

- Russian Mi-28s are currently being delivered without radar and ASE systems
- Chinese Z-10s currently have engine problems and there is little evidence of use of Z-10 weapons systems
- Eurocopter Tiger can only carry eight Hellfire missiles and does not have state-of-the-art FLIR.

Foreign attack helicopter platforms do have niche specialties:

- Chin-mounted cannons have attached ammo boxes that slew and elevate with a gun, a feature that tends to reduce jamming
- Chin-mounted cannons can be found on IFVs (for example, the 30-mm cannon on Mi-28 and BMP-3), which reduces maintenance training
- Non-U.S. attack helicopters tend to be:
  - Less expensive
  - Have variants compatible with Western avionics/weapons systems
  - Production contracts have in-country manufacturing agreements in order to remain competitive with U.S. FMS platforms
- Eurocopter has enhanced EMP protection.

The U.S. Army Apache and USMC Cobra can be compared:

- AH-64E:

  - Hellfire missile range out to eight km (using radar mode)
  - Dual-mode JAGM[41]
  - Radar capability
  - Level IV control over UAS (for payload and navigation)
  - Link-16 interoperability[42]

- AH-1Z:

  - 84 percent commonality with UH-1Y to reduce maintenance train
  - Software Reprogrammable Payload (SRP) and Variable Message Format (VMF) technologies (future digital interoperability improvements)[43]
  - Better targeting system:
    - Greater FLIR and Charge-Coupled Device Television (CCD-TV) identification ranges
    - Continuously variable zoom (versus static family of vehicles)
    - Lower spot jitter
    - Smaller laser beam divergence (more PGM accuracy at longer ranges)

- Both have tactical video data links and interoperability with UASs, ground combat systems (GCSs), and joint strike assets
- Both have state-of-the-art, fully integrated ASE suites.

---

[41] In February 2012, the Navy and the Marine Corps terminated its investment in the JAGM.

[42] Link 16 is tactical data link used to exchange near real-time communication, navigation, and identification information and supports information exchange between disparate C4I systems. The radio transmission and reception component of Link 16 is based on the high-capacity, ultra-high-frequency (UHF), line-of-sight waveform and can frequency hop, which provides secure, jam-resistant voice and digital data exchange. Link 16 operates on the principle of Time Division Multiple Access, wherein time slots are allocated among all network participants for the transmission and reception of data. Many joint Air-Ground platforms have Link 16 capability to include F-16s, F-15s, F/A-18s, E-2C2s, P-3Cs, EA-6Bs, EP-3Cs, RC-135s, KC-130Js, E-8J STARS, MH-60S/Rs, Carrier Battle Groups, Amphibious Ready Groups, Patriot Information Coordination Centers, USMC Tactical Air Command Centers, Air Force Air Operations Centers, and various UAS Ground Control Systems. Link 16 information is coded in binary J-series message protocol (MIL-STD 6016) and is passed in one of three data rates—31.6, 57.6, or 115.2 kilobits per second—all of which are relatively low bandwidth.

[43] VMF is bit-oriented digital information that facilitates the exchange of K-Series messages described in MIL-STD-6017 (Military Standard–6017). It is waveform independent, so it can be exchanged using wireless systems (such as HF, VHF, UHF, and SATCOM) and wired systems (such as Ethernet or fiber optic systems). K-Series message protocol enables it to have more-efficient use of bandwidth capacity due to the fact that multiple messages may use a single message heading. J-Series messages, however, require headers for each message, which decreases bandwidth capacity efficiency.

Medium-Lift Helicopters

UH-60M General Platform Information

The UH-60M Black Hawk (Figure 4.6) is a four-blade, twin-engine, medium-lift, utility helicopter manufactured by Sikorsky. It serves as the U.S. Army's utility helicopter for squad-size assault operations, but it also performs general support, command and control, and aeromedical evacuation missions. It was designed to replace the UH-60A and provides additional payload, range, and advanced digital avionics; better handling qualities; better pilot situational awareness; active vibration control; and improved survivability. Full-rate production began in 2007, and as of May 2013, the Army had approximately 400 in use. By FY 2027, the Army plans on procuring a total of 1,375 UH-60M Black Hawks through new builds.

Figure 4.6. UH-60M Black Hawk

SOURCE: Photo by the U.S. Army.

The UH-60M modernization program progresses within a block approach wherein Block I incorporates a digital cockpit, digital flight controls, wide-chord rotor blades, a more powerful General Electric T-700-GE-701D engine, and an integrated vehicle health management system (IVHMS). Block II incorporates improved ASE in the form of an enhanced AVR2B laser warning system; improved infrared suppression and the Common Missile Warning Systems (CMWS); improved survivability due to 23-mm antiaircraft artillery (AAA) tolerant rotor blades and a cockpit armed with Kevlar, glass fiber, and Nomex; and a Rockwell Collins Common Cockpit Avionics Architecture System (CAAS) that integrates communications, navigation, weapons, and mission subsystem information for improved pilot situational awareness.

Importantly, CAAS is compatible with K-Series VMF messaging and is installed on many other helicopter platforms, including a number of special operations forces helicopters (AH-6, MH-6, MH47D, MH-47E, and MH-60K), Army CH-47Fs, the Marine Corps CH-53E/Ks, and Coast Guard HH-60s.[44]

European EC725

The Eurocopter EC725 Caracal (Figure 4.7) is a five-blade, twin engine, glass-cockpit, medium-lift, multirole helicopter manufactured by Eurocopter. It is an evolved version of the SuperPuma/Cougar family of helicopters and is used by the French Air Force and Army for special operations, search and rescue, tactical transport, and aeromedical evacuation missions from the sea or on land. It was declared operational in May 2007, and by February 2013, 21 EC725s had been delivered to the French military. It has a 463 NM range using internal fuel tanks but can nearly double its range by carrying up to 630 gallons of auxiliary fuel.

Figure 4.7. Eurocopter EC725 Caracal

SOURCE: Photo by Jeff Web, CC BY-SA 3.0.

The Caracal improves on the SuperPuma/Cougar helicopter platforms by incorporating composite materials into the main rotor blades, a new rotor blade shape to reduce vibration levels, a fiberglass main rotor hub, armor plating to protect the cockpit and the main cabin,

---

[44] See *2014 Army Equipment Modernization Plan*, 2013; "UH-60M," *Globalsecurity.com*, last modified July 7, 2011; "Common Avionics Architecture System (CAAS)," *Rockwellcollins.com*, undated; Paul Clements and John Bergey, *The U.S. Army's Common Avionics Architecture System (CAAS) Product Line: A Case Study*, Carnegie Mellon Software Engineering Institute Technical Report, September 2005; David Jensen, "What's New with CAAS?" *Rotor & Wing Magazine*, October 1, 2010; *Sikorsky UH-60M Black Hawk Helicopter*, brochure, *Sikorsky.com*, July 2009; "UH-60M Evolution," PowerPoint presentation, undated; "Sikorsky S-70 (H-60) Upgrades," *Jane's Defence Equipment and Technology*, last posted January 23, 2013; "Sikorsky S-70A," *Jane's Defence Equipment and Technology*, February 7, 2013.

more-powerful Turbomeca Makila 2A1 turboshaft engines with full authority digital engine control (FADEC), a reinforced main rotor gearbox, an inflight refueling probe for extended mission range, an all-glass cockpit with liquid crystal multifunction displays, and an advanced helicopter cockpit and avionics system (AHCAS), and it can be fitted with a navigation FLIR and radar.

Medevac variants can carry up to 12 litters and combat assault support variants are armed with door-mounted 7.62-mm machine guns and side-mounted 68-mm rocket pods or side-mounted 20-mm cannon pods. All variants can be outfitted with a chaff/flare dispenser, radar warning receiver, and missile approach warning. The Caracal is also marketed for civilian firefighting purposes.[45]

Russian Mi-171A2

The Mi-171A2 (Figure 4.8) is a five-blade, twin engine, glass-cockpit, medium-lift, multirole helicopter manufactured by Mil Helicopters. Only two test aircraft currently exist, but the first deliveries are expected by the end of 2014. Currently marketed primarily as a civilian helicopter for export, the Mi-171A2 is designed to replace the Mi-8/17 "Hip" (which has been exported to nearly 60 countries) and features more than 80 technological improvements over its predecessor. These improvements include new, more powerful Klimov CK-2500PS-03 turboshaft engines with FADEC capability, a new rotor system with composite blades, a new airframe constructed with 20–30 percent composite materials, and the latest suite of integrated KBO-17 avionics developed by Radioelectronics Technologies, featuring an all-glass instrument panel and an onboard digital diagnostic maintenance system that monitors major component operating time while retaining information in the memory. The Mi-171A2 also features all-weather digital television and thermal imaging cameras for 360-degree situational awareness and collision avoidance, and a PKV-171 digital flight control system.

Figure 4.8. Mi-171A2

SOURCE: Photo by Doomych.

---

[45] "Eurocopter EC 225 and EC 725," *Jane's Defence Equipment and Technology*, February 1, 2013.

The Mi-171A2 has a 459-NM maximum range using internal fuel but can more than double its range by carrying up to 753 gallons of auxiliary fuel. It can accommodate 12 litters for the aeromedical evacuation mission and is touted for its potential civilian applications, such as forest protection, construction, loading and unloading operations, firefighting, and search and rescue. Although military applications for the Mi-171A2 are still in development, it is assessed that the new helicopter can carry up to 3,300 pounds of armament on wing stores that include antitank missiles, air-to-air missiles, rockets, and gun pods. ASE integration for this particular Mi-8/17 variant is also lacking, but older versions have been outfitted with the infrared jammers and suppressors, flare dispensers, and EW capabilities that are assessed to be readily transferable to the military variant of the Mi-171A2.[46]

General Characteristics of Medium-Lift Helicopters

Table 4.3 presents the general characteristics of medium-lift helicopters.

Table 4.3. General Characteristics of Medium-Lift Helicopters

Helicopter	Operational Introduction	Number in Service	Export Locations	Useful Payload	Max Speed (max/cruise)	Max Range/ Aerial Refuel
UH-60M Black Hawk	2007	1,192 (planned)	7 countries	11 combat loaded soldiers Int. capacity: 2,600 lbs. Underslung payload: 9,000 lbs	160/151 knots	276 NM Cannot aerial refuel Internal 360-gallon capacity
Eurocopter EC725 Caracal	2006	21 (current)	6 countries	25 seated combat troops Int. capacity: 5,500 lbs Underslung payload: 10,472 lbs	175/142 knots	463 NM Can aerial refuel Internal 683-gallon capacity
Mi-171A2 Hip	2014 (expected)	2 test aircraft exist Testing began in 2012, but fewer than 12,000 base models have been produced	None; however, nearly 2,500 base models have been exported to nearly 60 countries	26 combat troops Int. capacity: 8,818 lbs Underslung payload: 11,023 lbs	151/140 knots	459 NM Cannot aerial refuel Internal 898-gallon capacity

NOTE: Max range is without auxiliary fuel tanks.

---

[46] See "Mil Mi-17 (Mi-8M), Mi-19, Mi-171 and Mi-172," *Jane's Defence Equipment and Technology*, February 7, 2013; "Mi-171M: New Life of Venerable Helicopter," *Take-off Magazine*, July 2010; "Mi-171A2: Another Step Forward," *Russianhelicopters.com*, 2012; *Russianhelicopters.com*, undated; Rustechnologies, "Russia Presents Mi-171A2 Helicopter with New Avionics," *Ros Technologies Blog*, June 19, 2013; "Russian Helicopters Delivers First Mi-171A2 Fuselage," *Rianovosti*, January 23, 2012; Air Force Tactics, Techniques, and Procedures (AFTTP) 3-1, 2012, pp. 8-23–8-29; "Chapter Five: Russia and Eurasia," 2013.

Medium-Lift Helicopter Platform Trends and Insights

The section presents trends and insights regarding medium-lift helicopter platforms. There is increased sophistication in U.S. *and* rest of the world (ROW) platforms. All medium-lift helicopters are being upgraded or built with:

- Composite materials
- Enhanced survivability
- More-fuel-efficient engines with FADEC capability
- FWB electronic controls
- All-glass cockpits with large-screen, liquid-crystal, multifunctional displays
- Multipurpose flexibility—e.g., FLIR *and* weapons options.

U.S. Army UH-60M Black Hawks have a smaller useful payload, but it is purposefully built for U.S. Army squad operations. Black Hawks also have an advantage in:

- Digital network connectivity and interoperability with joint air-ground platforms
- Aircraft survivability.

Foreign platforms have some niche specialties (and different design philosophies):

- Mi-171A2 has robust armament capabilities
- EC725 can conduct aerial refueling
- EC725 can perform dedicated combat search and rescue (CSAR) mission using dedicated radar capabilities
- All variants of EC725 and Mi-171A2 have navigation FLIRs
- Foreign systems have higher useful payloads and medevac capacities
- Foreign systems have longer maximum range using internal and auxiliary fuel tanks.

Heavy-Lift Helicopter Platforms

CH-47F General Platform Information

The CH-47F Chinook (Figure 4.9) is a multiengine, tandem-rotor, glass-cockpit, heavy-lift helicopter manufactured by Boeing. It is employed in the U.S. Army's general support aviation battalions (GSABs) and conducts troop movement, artillery emplacement, and battlefield resupply missions. As of November 2012, the Army had received 211 F-model Chinooks, but the Army Modernization Program calls for a total of 464 CH-47Fs, through a combination of 253 complete new builds and 211 like-new aircraft using selected recapitalized CH-47D components, by 2017. These deliveries occur within two multiyear contracts, known as MYI and MYII.

Figure 4.9. CH-47F Chinook

SOURCE: Photo by Sgt. 1st Class Roy Henry, Public Affairs Office, Georgia Department of Defense, CC BY 2.0.

The CH-47F incorporates several new technologies, including a new "machined" airframe that decreases weight and increases strength, the more powerful and fuel-efficient Honeywell T55- GA-714A turboshaft engines with FADEC capability, the same Rockwell Collins CAAS cockpit installed in the UH-60M that has digital data bus architecture, high-definition liquid crystal multifunction displays, redundant high-integrity Ethernet data buses, electronic K-Series VMF messaging, built-in systems diagnostics, and modularity for future upgrades. The F-model Chinook also features the Digital Advanced Flight Control System (DAFCS), which provides a level of flight automation to typical maneuvers performed by Chinook pilots, such as hovering and landing from a hover.

Future planned upgrades also include the composite Advanced Chinook Rotor Blade (ACRB) that will debut in 2016 and increase aircraft lift capacity by 2,000 pounds; the Active Parallel Actuator System (APAS) to enhance the DAFCS by providing improved rotor torque management; a new streamlined and more efficient fuel system (derived from the special operations MH-47G variant) that decreases fuel system weight and adds fuel capacity, thereby increasing range and lift capacity; and a completely upgraded electrical system.[47]

---

[47] See *2014 Army Equipment Modernization Plan*, 2013; Sofia Bledsoe, "Team Chinook Signs CH-47F MYII Contract; Cost Savings of $810 Million," *Army.mil*, June 14, 2013; Graham Warwick, "Block 2 CH-47F to Tackle Payload Shortfalls," *Military.com*, April 22, 2013; "Boeing Awarded U.S. Army Contract for 14 Additional CH-47 Chinook Helicopters," news release, *Boeing.com*, January 11, 2012; "CH-47F Chinook Backgrounder," *Boeing.com*, March 2012; "CH-47F Improved Cargo Helicopter (ICH)," *Globalsecurity.org*, July 2011; Scott Gourley, "Aviation Modernization Milestone Update," *Army Magazine*, January 2013; Joakim Kasper Oestergaard, *Boeing CH-47 Chinook*, Aerospace and Defense Intelligence Report, November 30, 2012.

USMC CH-53K

The CH-53K King Stallion (Figure 4.10) is a seven-blade, three-engine, heavy-lift cargo helicopter manufactured by Sikorsky. It is currently in development, and the first preproduction test aircraft is scheduled for delivery in late 2013. It is scheduled to become operational in 2018 in the USMC, and USMC procurement plans call for a total of 227 CH-53Ks. When it is fielded, the CH-53K will be the largest and heaviest helicopter in the U.S. military, at a maximum gross weight of 88,000 pounds.

Figure 4.10. CH-53K King Stallion

SOURCE: Publicity photo from Sikorsky.

The CH-53K improves on the CH-53E aircraft currently in use by the U.S. Marine Corps and features more-powerful General Electric GE38-1B engines, high-efficiency composite rotor blades with anhedral tips, and a composite airframe that significantly increases useful payload. The CH-53K also has a wider cabin to fit a High Mobility Multipurpose Wheeled Vehicle (HMMWV) internally, fly-by-wire electronic flight controls, a navigation FLIR with integrated HMS/D, and a significantly revamped cockpit that features the Rockwell Collins Avionics Management System (AMS), which, like the UH-60M and CH-47F, integrates communications, navigation, weapons, and mission subsystem information for improved pilot situational awareness and is compatible with K-Series VMF messaging.

Survivability improvements also include advanced lightweight armor protection for the aircrew and crash-worthy seats for troops in the cabin. The CH-53K will be able to carry 24 litters in the aeromedical evacuation role, will also be able to carry two 463L pallets internally, and has three cargo hooks that can carry an external load of 36,000 pounds.[48]

---

[48] See U.S. Government Accountability Office, *Defense Acquisitions: Assessments of Selected Weapon Programs*, GAO-13-294SP, March 2013, pp. 53–54; "Marines Up Order for New Heavy Lifter," *Rotor & Wing*, August 1, 2007; "CH-53X Super Stallion," *Globalsecurity.org*, undated; "New Heavy Lift Helicopter Starts

Mi-26T2

The Mi-26T2 (Figure 4.11) is an eight-blade, twin-engine, heavy-lift helicopter manufactured by Mil Helicopters. It is currently in development and is planned to be used by the Russian Air Force after its scheduled operational debut in 2014. The Mi-26T2 is a modernized version of the Mi-26 Halo, which is the biggest helicopter in the world, and has several updated technologies, including glass-fiber reinforced plastic (GFRP) rotor blades, a titanium rotor head, and Ivchenko-Progress ZMKB D-136-2 engines with FADEC capability that can produce 11,490 shaft horsepower.

Figure 4.11. Mi-26T2

SOURCE: "Mil Mi-26T2" shared via Flickr by José Luis Celada Euba, CC BY 2.0.

The Mi-26T2 also features a modern avionics suite designed by Ramenskoye Design Company, including five large-screen multifunctional LCDs, a digital navigation system, FWB electronic controls, an integrated navigation FLIR and laser rangefinder, and a closed-circuit television camera to observe slung payloads.

The cargo load of the Mi-26T2 is wider and taller than a C-130J and is capable of carrying two airborne infantry combat vehicles, 90 combat-loaded troops, 60 aeromedical evacuation litters, or six 463L pallets. The aircraft is also being marketed for civilian purposes, such as construction, logging operations, and firefighting duties.[49]

Development," U.S. Marine Corps, press release, January 9, 2006; "Sikorsky Aircraft Selects Rockwell Collins to Provide CH-53K Avionics Management System," *Sikorsky.com*, June 29, 2006; "Sikorsky CH-53K Super Stallion," *Jane's Defence Equipment and Technology*, August 3, 2012; *Sikorsky CH-53K Helicopter*, brochure, *Sikorsky.com*, June 2007.

[49] See "Mil Mi-26 and Mi-27," *Jane's Defence Equipment and Technology*, June 24, 2013; "Mi-26T2," *Deagel.com*, August 11, 2011; "Mass Production of Mi-26T2 Will Begin in 2012 Year," *Aviationunion.org*, January 3, 2012.

General Characteristics of Heavy-Lift Helicopters

Table 4.4 presents the general characteristics of heavy-lift helicopters.

Table 4.4. General Characteristics of Heavy-Lift Helicopters

Helicopter	Date of Introduction	Number in Service	Useful Payload	Cargo Bay Dimensions	Max Speed (max, cruise)	Max Range/ Aerial Refuel
CH-47F Chinook	2007	464 (planned)	33 combat soldiers Int. capacity: 28,000 lbs Ext. capacity: 25,000 lbs	30.5-ft. long 7.5-ft. wide 6.5-ft. tall	170/130 knots	400 NM Cannot aerial refuel Int. fuel: 1,034 gallons
CH-53K King Stallion	2015 (expected)	227 (planned)	37 combat troops Int. capacity: 35,000 lbs Ext. capacity: 36,000 lbs	30-ft. long 9-ft. wide 6.5-ft. tall	170/140 knots	454 NM Aerial refuelable Int. fuel: 2,286 lbs
Mi-26T2 Halo	2014 (expected)	42 (planned)	90 combat troops Int. capacity: 44,092 lbs	39ft 7.5" long 8' wide 10' 4.75" tall **wider and taller than C-130J**	159/132 knots	430 NM Cannot aerial refuel Int. fuel: 3,170 gallons

NOTE: Max range is without auxiliary fuel tanks.

Heavy-Lift Helicopter Platform Trends and Insights

The section presents trends and insights regarding medium-lift helicopter platforms. There is increased sophistication in U.S. and ROW platforms. All heavy lift helicopters are being upgraded or built with:

- Composite materials
- Enhanced survivability
- More-fuel-efficient engines with FADEC capability
- FWB electronic controls
- All-glass cockpits with large-screen, liquid-crystal, multifunctional displays.

U.S. heavy lift helicopters have a significantly lower payload than Mi-26T2, yet greater maneuverability. They also have a greater digital network connectivity than Mi-26T2.

Regarding foreign niche specialties, Mi-26T2 has many variants (flying hospital, crane, tanker, and firefighter), which makes it more marketable for civilian purposes. And Mi-26T2 is truly in a different cargo class (C-130 class payload).

"Mi-26 T2: Multipurpose Transport Helicopter," undated; "Mi-26T2 Versus CHINOOK," *Take-off Magazine*, February 2011; "New Mi-34C1, Ka-226T, Mi-38, Mi-26T2 Showcased at MAKS 2011," *Russianhelicopter.aero.en*, August 6, 2011; "Rostvertol Will Demonstrate the Modernized Mi-26T2 Heavy Transport Helicopter to Algerian Air Forces," *RussianAviation.com*, June 20, 2012; "Chapter Five: Russia and Eurasia," 2013.

If we compare the CH-47F with the CH-53K, we find:

- The CH-47F is much less expensive to operate ($12,000 per flight hour versus $20,000)
- The CH-47F is more maneuverable and can access more-confined landing zones due to tandem rotor and faster approach speeds
- The CH-47F has structural modifications for faster loading and unloading on C-5
- The CH-53K has higher payload capability:
  - The CH-53K has a third engine and can carry 7,000 pounds more.
  - The CH-53K is 18 inches wider, which can accommodate HMMWVs and Mine Resistant Ambush Protected (MRAP) vehicles
- The CH-53K has FLIR and integrated helmet-mounted navigation system
- Both have cutting-edge tactical digital connectivity
- Both have extensive avionics upgrades.
- Both have similar strategic transportability (two on C-5)
- Both can carry three 463L pallets.

Scout Helicopters

Recent research into global trends at the unclassified level in scout and reconnaissance helicopters indicates that very few new specialized observation helicopters are being developed or procured. In particular, The International Institute for Strategic Studies 2013 *Military Balance* journal shows limited proliferation and procurement of scout helicopters as well as a reduction in scout helicopter inventories. France and Australia, for example, are replacing their aging scout helicopters (Gazelles and OH-58 Kiowas, respectively) with attack helicopters (Eurocopter Tigers) that are to be used in the dual roles of attack and reconnaissance. They are also procuring more UAS assets designed to be interoperable with these attack helicopters. Thus, U.S. Army plans to procure a new observation helicopter are an outlier in terms of global trends.

It should be noted that the U.S. Army also uses its OH-58 aircraft in a light-armed attack role. In Iraq and Afghanistan, for example, OH-58s provided armed reconnaissance for convoys, trying to identify ambushes prior to a convoy reaching the danger point. Those same helicopters provided fires when troops were in contact with the enemy. So in that sense the Army is using the OH-58 aircraft in a light-attack role, which is beyond the helicopter's original mission profile.

Implications from this observation are varied. On the one hand, it may be concluded that the *need* for a dedicated scout helicopter platform is decreasing given the fact that many scout helicopter missions (not all), *can* be performed by a mix of UAS and reconnaissance-capable attack helicopters. On the other hand, these procurement decisions may derive from constraints on military spending and the need to reduce training, maintenance, and logistics support associated with disparate platforms more so than from operational demands. More analysis is needed to determine what is best for the U.S. Army in this regard; however, global trends are not congruent with U.S. Army plans to procure a new observation helicopter.

(This page is intentionally blank.)

5. Logistics and Engineering

Over the past decade, U.S. Army combat and logistics vehicle development has focused on responding to the pervasive IED threat encountered in Iraq, Afghanistan, and elsewhere. The response to the threat has included drastically increased protection and, consequently, drastically increased size and weight ground vehicles. The increase in protection, size, and weight of tactical ground vehicles has influenced other existing support vehicles, such as the landing and amphibious craft used to deliver ground vehicles. The research into the trends in logistics vehicles of other armies clearly shows that they also perceive the increased threat from mines and IEDs and are adding considerable amounts of armor to their fleets of supply vehicles.

While these developments have improved the survivability of U.S. forces, these changes have resulted in a considerable increase in the size and weight of the Army's logistics vehicles. This discussion and analysis specifically focuses on the platform and supporting technology developments for logistics vehicles, engineering vehicles, and amphibious delivery platforms. The systems included in this chapter are the result of specific requests by the sponsor.

Trends in Logistics and Other Ground Support Vehicles

The pervasiveness of asymmetric and dispersed threats, such as IEDs, in contemporary combat operations has forced almost all militaries to consider the impact of these threats on tactical platforms. While initial efforts primarily focused on improving vehicles intended for close combat, the growing need for logistics vehicles to fight through these asymmetric threats on their way to their destination has motivated significant development of protection alternatives for logistics and other support vehicles. Often without infantry support, these platforms are required to provide their own security and prepare for direct contact with the enemy. Additionally, recent operations have demonstrated the opportunities for logistics vehicles to conduct other functions, such as surveillance or reconnaissance, as the convoy moves to its destination. These factors have motivated pursuit of logistics and other support vehicles that are both deployable and protected with heavier weaponry, more sensors, and better communications.[50]

---

[50] Francis Tusa, "Wagon Train: Logistics Lessons from Operations," *Jane's International Defence Review*, November 7, 2012.

Protection

Initial attempts at improving the protection of logistics and other support vehicles, like the efforts for combat vehicles, have primarily included hard mounting additional armor and building vehicles with integrated and welded armor packages.[51] However, this approach has resulted in massive vehicles, such as Navistar's MRAP Wolfhound, a flatbed variant of the Cougar MRAP employed by U.S. forces.

Because of their weight (often in excess of 25 tons), these vehicles are not readily deployable and provide little flexibility to tailor armor to operations demanding less protection and more mobility or transportability.

Some countries have addressed the vehicle weight and transportability issues by pursuing strategies such as procuring a smaller fleet of well-armored vehicles for high-threat environments and a larger fleet of soft-skin logistics vehicles for more permissive applications.[52] To develop a more readily deployable logistics support fleet, countries such as the United Kingdom and France are investing in families of logistics vehicles that can be tailored with varying levels of armor protection.[53] These vehicles are designed to enable the rapid addition of protection through integrated armor packages that can be quickly added (or removed). This approach enables relatively easy tailoring of armor packages based on the specific local threat conditions, while minimizing the base platform weight for easier deployment and sustainment.

The add-on armor approach is also enabling militaries to replace armor panels as material technologies improve. For example, the AmSafe Bridport Company and the UK Defence Science and Technology Laboratory (DSTL) have partnered to develop the Tarian lightweight fabric armor, which is 85 percent lighter than steel armor and half the weight of aluminum systems, allowing for quick application to and removal from vehicles.[54] The British Tarian armor system is claimed to be more effective than heavier bar armor in protecting against rocket-propelled grenades (RPGs). As illustrated in Figure 5.1, armor panels are secured to the vehicle with quick release devices at each corner, enabling the rapid replacement of damaged panels. This material provides a relatively light and tailorable method to realize significant increases in blast protection.

---

[51] "Executive Overview: Logistics Support and Unmanned Vehicle Technology," *Land Warfare Platforms: Logistics, Support & Unmanned*, March 20, 2014.

[52] "Executive Overview," 2014.

[53] British Land Forces, *British Army: Vehicles and Equipment*, United Kingdom Ministry of Defence, 2012.

[54] "Fact Sheet: Research and Development of Tarian," Defence Science & Technology Laboratory, United Kingdom Ministry of Defence, 2012.

Figure 5.1. Tarian Protective Fabric on the British HET (left) and Spartan (right) Vehicles

SOURCE: Photos shared by *Think Defence* via Flickr, CC BY-NC 2.0.

Protection capabilities have also been added to other support platforms that are likely to be exposed to lethal enemy fires, such as armored bulldozers. As pictured in Figure 5.2, the Israeli Defense Force (IDF) has fielded a fully armored D9R bulldozer that is significantly used by combat engineers in urban combat operations to deal with mines, roadside bombs, and other area denial threats.[55] These platforms are also fitted with front and back sensors to give the operator better situational awareness. Additionally, Israel Aircraft Industries has developed a remote control kit for the D9 dozer to remove the human occupant in the most dangerous situations.

Figure 5.2. Israeli Defense Force Caterpillar D9R

SOURCE: Photo by MathKnight, CC BY 3.0.

---

[55] "Israel Military Industries Bulldozer Protection Kit," *Jane's Defense & Security Analysis*, July 13, 2012.

Countermeasures

While a number of classified countermeasure technologies are being pursued by the U.S. military and the defense industry, other emerging technologies are more widely marketed and available for purchase. For example, the Cassidian CPJ COMPACT-R German IED countermeasure system applies smart responsive jamming technology that detects and classifies in the 20-MHz to 6-GHz frequency.[56] This system detects the signal frequency and responds by transmitting real-time jamming signals tailored to the detected hostile frequency band. This system increases power efficiency and decreases the potential for inadvertent jamming by focusing only on the necessary signal frequency. The COMPACT-R and other electronic countermeasure systems are pursuing more-responsive and targeted jamming in order to reduce the unwanted externalities produced by the original wholesale jammers that often interfered with friendly convoy communications and other electronic equipment.

Autonomous Vehicle Technologies

Current investment in driver-assist technology has been primarily led by civilian initiatives, such as Google Cars (United States), Volvo Truck Train (Europe), New Energy and Industrial Technology Development (Japan), and Guardium (Israel). While only the U.S. military has established a formal driver-assist procurement program, a number of countries are testing appliqué kits and new vehicles that enable human, remote, or fully autonomous control.[57] The appliqué kits for driver-assist technologies are the most promising, allowing the addition of driver-assist technology to existing platforms at one-tenth of the cost of integrating technology into a new platform. The U.S. Army is interested in this technology, and several other armies are also exploring the possibilities. For example, interviews with British Army logistics officers indicate that their army is moving in the direction of driver-assist technologies, including autonomous vehicles. This is still an area of emerging civilian technology that should be monitored for military use.

Other Emerging Technologies with Potential Benefits for Logistics and Support Vehicles

In addition to protection, countermeasure, and autonomous-vehicle technologies, other countries are pursuing a broad range of technologies that can improve the performance and safety of logistics and other support vehicles. The United Kingdom's Ministry of Defence (MOD) Future Protected Vehicle (FPV) program has developed seven concept vehicles to highlight a host of new technologies. This program has also demonstrated a total of 47

---

[56] "CPJ COMPACT-R Convoy Protection Jammer with Smart Responsive Jamming Technology," *Jane's Defense & Security Analysis*, March 4, 2013.

[57] Brian Lesiak, email correspondence, April 1, 2013.

emerging technologies identified as potentially suitable for development for combat and support vehicles, including:[58]

- An active visual management system that projects imagery from behind a vehicle onto its front, theoretically rendering it invisible
- 3-D imaging for situational awareness
- Unattended remote acoustic sensors
- An engine with an integrated starter generator
- A nonmechanical thermal management system that uses phase-change materials to exploit the heat from the vehicle's exhaust to manipulate its signature in the infrared spectrum
- New armor concepts that include lightweight and "transparent" solutions
- A nonmechanical "butterfly" armor package that uses electromagnetic systems to lift composite armor panels and create an air gap between two plates, replicating spaced armor but without its bulk.

Key Logistic and Support Vehicle Development Observations

Due to the pervasive and continued asymmetric warfare threats to logistics and other support activities, the need to increase protection, increase armor, and reduce the potential for casualties remains. However, numerous countries are also pursuing ways to limit the impact of increased armor on the transportability of these platforms. A review of these varying approaches supports the following observations:

- **Increasing armor protection and integrity are still the predominant methods for improving the survivability of logistics and support elements.** While these efforts have previously resulted in a significant increase in vehicle weight, modular armor concepts with new materials are providing countries with more tailorable options to meet operation-specific threats.
- **To provide a flexible range of options and achieve a more cost-effective fleet composition, many countries are going with a mixed logistics and support vehicle fleet.** These mixed fleets often include:
  - Commercial off-the-shelf (COTS) platforms
  - Tactical unarmored vehicles
  - Factory-built, welded armored cabs that provide complete cab enclosure for MRAP vehicles
  - Rapid up-fit protection suites that allow unarmored base platforms to accommodate various levels of armor as required by operations.
- **With the increased armament and protection, support vehicle are being relied on more to conduct patrol and collection tasks while providing their own security.** These additional tasks have required current and emerging

---

[58] Huw Williams, "Future Protected Vehicle Study Turns up Host of Concepts and New Technologies," *Jane's International Defence Review*, January 11, 2001.

support vehicles to accommodate improved communications, armament, countermeasures, and situational awareness technologies.

- **The U.S. Army could benefit from taking a mixed-fleet approach to its logistics vehicles and support vehicles.** This approach would facilitate tailoring the vehicles based on the specific operation and threat. Highly protected logistics vehicles may be needed in some situations, but not all.

Trends in Amphibious and Riverine Delivery Platforms

While a broad range of countries maintains some amphibious and riverine delivery capabilities, overall global investment in and development of these capabilities have been relatively limited and generally stagnant, with only 4 percent and 2 percent of world naval investment programs for landing craft and riverine craft, respectively.[59] Due to this general lack of interest and competition in the landing craft and riverine craft markets, the available capabilities for potential use have changed little in the last two decades. For example, current Landing Craft Utility (LCU) vehicles generally have top speeds of up to 20 knots, representing little improvement over craft available 40 years ago. The U.S. Navy's and Landing Craft Air Cushion (LCAC) is a hovercraft that is capable of much faster speeds, up to roughly 40 knots if the sea conditions permit. Currently, the United States relies on LCU and LCAC platforms that have been in service for more than 25 years. Additionally, these existing craft were designed to transport ground vehicles that were much lighter than the current platforms. The existing LCU and LCAC platforms can accommodate the size and weight of a current four-vehicle Stryker platoon.[60] The criterion of being able to lift a Stryker platoon was a key consideration of the DAMO-FD sponsor of this portion of the research.

Current Leading Landing Craft Utility Platforms

While global investment in LCU platforms has been limited and generally stagnant, newly conceived and initiated programs are promising improved alternatives in the near future. U.S. LCU capabilities have changed very little since the LCU-1600 entered service in the 1970s. The LCU-1600 can accommodate the weight of a Stryker platoon and the length of four Stryker vehicles in line. However, the LCU-1600's limited top speed of 11 knots is a limitation.

---

[59] "Naval Construction, Naval Forecast and Naval Upgrades," Carpenter Data Partnership (CDP) Group, 2011.

[60] "Stryker," U.S. Army Fact Files, *Army.mil*, undated. The current double v-hull variant of the Stryker weighs approximately 38,000 pounds, making the total weight of a four-vehicle Stryker platoon approximately 160,000 pounds. While the current LCU has sufficient capacity to carry a Stryker platoon at an approach speed of 11 knots, the LCAC cannot accommodate the full size or weight of a Stryker platoon. For this analysis, the Stryker platoon is used for consideration due its role as the smallest contained mechanized maneuver element for forced entry and early entry operations.

Figure 5.3. Current U.S. LCU-2000

SOURCE: Photo by the U.S. Navy.

Taking the place of the older LCU-1600 is the LCU-2000 *Runnymede* class of utility landing craft (see Figure 5.3). This class, built in the 1990s, has a speed of 12 knots and can lift up to 350 tons of cargo, including fighting vehicles.[61]

While most other countries have maintained similarly limited LCU capabilities, there are LCU platforms in development that promise to drastically improve LCU performance potential. Of the most impressive potential LCU programs, only the United Kingdom's PASCAT has made it to demonstration. The PASCAT, like many emerging alternatives, utilizes a novel catamaran hull form to reduce draft and increase maximum speed.[62] The PASCAT requires fewer than three meters of draft when operating in catamaran mode. The PASCAT would provide almost a 75 percent increase in top speed over the U.S. LCU-1600. While other programs promise even better carry capacity and top speed, these programs are still in the planning stages and have not been demonstrated as yet. As with the LCAC platforms, the general trend in LCUs is the inclusion of more armament and countermeasures, such as air defense systems, to negotiate less permissive environments.

Current Leading Landing Craft Air Cushion Platforms

While many countries have invested in LCAC, or hovercraft, platforms because of their ability to operate with fewer than six feet of draft, most of these platforms have been designed to

---

[61] "Landing Craft Utility (LCU)," *GlobalSecurity.org*, undated.

[62] Richard Scott, "Race to the Beach: Novel Hullforms Push the Pace," *Jane's Navy International*, April 20, 2011.

transport personnel and light cargo rather than one or more armored vehicles. However, the United States, Russia, and China have each developed LCAC platforms capable of carrying varying numbers of larger vehicles. Only the massive Russian-made Pomornik is capable of carrying a four-vehicle Stryker platoon. Additionally, foreign LCAC platforms are trending toward an increase in armament in order to negotiate more-formidable marine and airborne threats. With no new large-scale hovercraft programs in production, opportunities for the existing LCAC in the near-term are likely limited to incremental improvement through modernization.[63] While the U.S. Navy has commissioned designs for the Ship-to-Shore Connector (SSC) that, as depicted in Figure 5.4, will replace the current LCAC and will accommodate slightly less weight than a full Stryker platoon.

Figure 5.4. Planned Characteristics of the U.S. Navy Concept Ship-to-Shore Connector

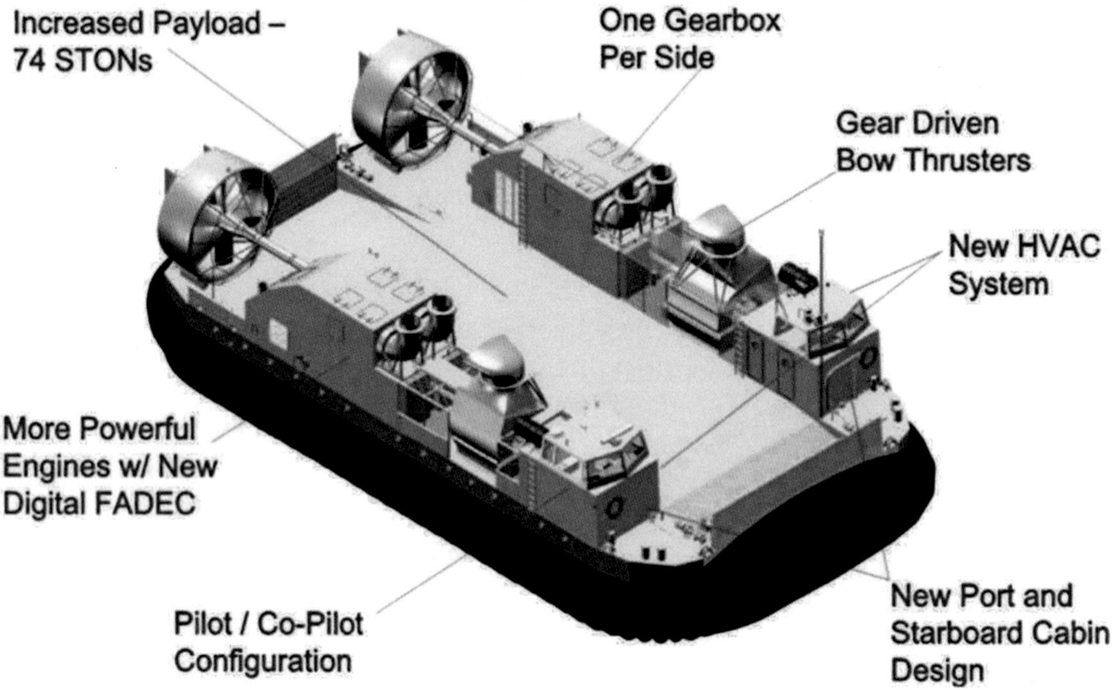

SOURCE: Photo by the U.S. Navy.

---

[63] However, in March 2011 the U.S. Navy released a draft request for proposals for a new SSC to gradually replace the LCAC fleet for planned delivery in FY 2018.

Key LCAC and LCU Development Observations

This survey of existing and emerging options for the potential delivery of U.S. Army forces during riverine and amphibious operations supports some general observations that should inform the Army's consideration of capability development options:

- **Due to the need for landing craft that can enable immediate roll-on and roll-off of Army ground forces, such as a four-vehicle Stryker platoon, landing craft are the only suitable marine options.** The likely need to rapidly move and off-load more than 150 tons of equipment and personnel as part of amphibious or riverine entry operations will limit marine platform options almost exclusively to LCAC, LCU, and Landing Craft Tank (LCT) platforms.
- **While some variations in hull shapes are being explored for LCU and LCT platforms, the available capabilities have improved little in recent decades.** With only 6 percent of world naval procurement pursuing LCAC, LCU, and LCT platforms, there is little pressure from the world market to motivate development and adaptation of emerging technologies to improve amphibious or marine delivery capabilities, especially for loads over 20 tons.
- **Within the LCAC class of platforms, only the Russian-made Zubr-class hovercraft has sufficient payload capacity to deliver a four-vehicle Stryker platoon.** Due to the technological limitations associated with hovercraft platforms, there are few options with more than 50 tons of carrying capacity and only one current option with more than 100 tons of carrying capacity. While the LCU and LCT platforms can often accommodate 150 tons or more, their maximum speed and draft limitations will challenge their utility for rapid delivery of a Stryker platoon or similar mounted formation as part of dynamic amphibious and riverine operations.
- **New LCU development and demonstration programs offer promising but unverified potential for increases in speed, accessibility, and carrying capacity for LCUs and LCTs.** A few current programs are leveraging innovative catamaran tri-hull designs that promise to dramatically increase approach speed. However, only the United Kingdom's PASCAT LCU platform has been demonstrated, and it has a planned carrying capacity of 80 tons, well short of the capacity required to deliver a Stryker platoon.

(This page is intentionally blank.)

6. Protection

This chapter is devoted to the protection warfighting function. This is a potentially huge issue that could include such disparate topics as air and missile defense, body armor for individual soldiers, protection of logistics and other vehicles from the ever-more-common IED threat, and many other areas. Due to the time and space limitations of the project, this chapter focuses on the defense against artillery, rockets, mortars, and unmanned aerial systems. That said, some of the systems that will be described here have applicability in defense against cruise and short-range ballistic missiles.

One of the reasons for the focus on the artillery and rocket challenge is due to the growing threat posed by long-range rocket systems, which was highlighted in Chapter Three. As the range of multiple rocket launchers has increased in the past decade, those systems are now capable of firing at distances that were once the purview of short-range ballistic missiles, such as the Chinese WS-2, which has a range of more than 200 km. Multiple rocket launchers are easy-to-hide, highly mobile systems that tend to fire in salvos. Therefore, defensive systems could be confronted with an incoming barrage that could range from half a dozen to literally scores of weapons. This has several implications. For example, while the radar of the Army's Patriot missile defense system is capable of tracking an incoming barrage of long-range rockets, and Patriot interceptors could be launched against the incoming threat, the cost-effectiveness of this option is decisively against the defender. The latest version of Patriot costs more than $3 million per missile, making it prohibitively expensive for use against barrages of artillery rockets.

One of the major challenges associated with today's long-range rockets is their high approach velocity. Whereas the older 122-mm Russian-made Katushya rockets with ranges of 20–40 km (which have frequently been fired into Israel from southern Lebanon) travel at speeds of roughly 700 meters per second, long-range, heavy MRL rockets such as WS-2 are much faster, on the order of 1,800 meters per second.[64] The much higher speeds of long-range artillery rockets make them a more difficult challenge for defensive systems. For example, whereas the Israeli Iron Dome counterrocket and -artillery system can cope with the older Katushya rockets, the current Israeli system would have much greater difficulty countering the higher-speed Chinese WS-2.

The remainder of the chapter examines a number of foreign air defense systems that might be useful for protection against the rocket threat, as well as the growing number of UASs. Additionally, some U.S. Navy systems that might be useful for defense of ground units or targets are also included in the comparison. Since the U.S. Army would probably place a high premium on systems that could be quickly deployed into an operational area, special attention was placed

---

[64] "Katushya Rocket," *GlobalSecurity.org*, undated.

on identifying foreign systems that appeared to be relatively easy to deploy. The foreign defensive systems include guns, missile, and speed-of-light weapons.

Gun Defensive Systems

Guns as point-defense systems have the advantage of being relatively small and self-contained compared with missile systems. Some gun systems are mobile, being mounted on tracked or wheeled chassis. Others, such as the U.S. Navy's close-in weapons System (CIWS) that has been modified for ground use in Iraq and Afghanistan, are fixed systems mounted on pedestals. Indeed, modified versions of the Navy CIWS became the Army's main counter rocket, artillery, and mortar (C-RAM) capability in Iraq. What follows are a number of foreign gun systems that are used in a point-defense role against rockets, artillery, mortars, and UASs.

Skyshield (Figure 6.1) is a short-range 35-mm gun system developed by the Swiss company Oerlikon. A radar-directed system, this weapon can be used against aerial targets in all weather conditions. Its high-speed projectile (roughly 1,400 meters per second) explodes in midair, creating a cloud of fragments that destroy or damage incoming projectiles or UASs. The German Army currently uses Skyshield as its point-defense system for high-value targets. Israel has also examined the system as a way of adding additional capability to its defense against rockets.[65]

Figure 6.1. 35-mm Oerlikon Skyshield System

SOURCE: Publicity photo from Rheinmetall.

A mobile version of Skyshield has been developed, the Boxer system shown in Figure 6.2. The advantage of this system is its relative ease of deployment and low cost compared with a

---

[65] Christopher F. Foss, "Skyshield Can Fire AHEAD," *Israeli Homeland Security*, February 20, 2013.

missile defensive system. It is not clear, however, how well the system would perform against a high-speed artillery rocket such as WS-2.

Figure 6.2. Boxer Mobile Variant of the Skyshield System

SOURCE: Publicity photo from Rheinmetall.

Another gun defensive system is the Italian-made Oto Melara Draco air defense system (Figure 6.3). Whereas most defensive gun systems are in the 20–40-mm class, this Italian system uses a high- velocity 76-mm weapon that has a much greater ability to engage other targets, such as armored vehicles and buildings, compared with the lighter-caliber cannons. Its range of 8 km against aerial targets is roughly double that of 40-mm weapons. The rate of fire is roughly 80–100 rounds per minute, considerably slower than smaller-caliber automatic cannons.

Figure 6.3. Italian 76-mm Draco System

SOURCE: Publicity photo from OTO Melara.

This system would have considerable utility against UASs (or helicopters), but would be challenged if expected to engage high-speed artillery rockets. The mobility of the system is impressive, and it is an example of how defensive gun systems can be vehicle mounted.[66]

Another foreign gun system is the Chinese-made LD-2000 30-mm gun (Figure 6.4). This truck-trailer-mounted weapon has a very high rate of fire, more than 4,000 rounds per minute, and a maximum range of roughly 3,000 meters. While it is well suited to engage UASs and slower rockets, it would have difficulty in engaging high-speed artillery rockets, particularly if they eject submunitions before entering the gun's engagement range. The mobility of the system is, however, impressive. This weapon is being marketed by the Chinese NORINCO company.[67]

Figure 6.4. Chinese LD-2000

SOURCE: Publicity photo from Norinco.

Guns all have the advantage of being cheaper and smaller when compared with missile defensive systems. They are often more mobile, which can be an important consideration for rapidly moving ground forces, particularly ones that have to deploy into an operational area, as is the norm for U.S. forces. Although they are good against UASs and slower rockets (as well as mortar rounds, which are very-low-velocity weapons compared with rockets and cannon projectiles), guns would be hard pressed to cope with high-speed artillery rockets. Additionally, due to their short range, guns are truly point-defense systems that cannot compare with defensive missile systems in their ability to engage approaching threats at greater distances.

For comparison purposes, the U.S. Navy's CIWS (often referred to by its nickname, "Phalanx") was first deployed aboard Navy combat ships in 1980 and support vessels starting in 1984 (see Figure 6.5). The system is a 20-mm cannon with a very high rate of fire, roughly 4,500 rounds per minute, and an effective range of about 3.5 km. The short range of the CIWS means

---

[66] "Italian 76mm *Draco* System," *Military-Today.com*, undated.

[67] Carlo Kopp, *Russian and PLA Point Defence System Vehicles*, *Air Power Australia*, June 2008, updated April 2012.

that it is truly a point-defense system; ships armed with Phalanx cannot protect other vessels unless they are very close.

Figure 6.5. U.S. Navy CIWS

SOURCE: Photo by the U.S. Navy.

The CIWS was used in Iraq, mounted on tall concrete pedestals, to protect fixed bases from incoming rockets and mortar fire. From the U.S. Army's perspective, this system has the advantage of being part of the U.S. inventory.[68]

Defensive Missile Systems

In recent years there has been increasing interest in defensive missile systems as the threat from ballistic missiles, cruise missiles, and UAS has increased. The American Patriot system was first used in a missile defense mode in the 1991 Persian Gulf War. A number of militaries around the world have developed defensive missile systems, or are in the process of doing so. As the threat of artillery rockets has increased, the use of defensive missiles against that threat is also increasing, as evidenced by the Israeli Iron Dome system (Figure 6.6).

The advantage missiles have compared to guns is their ability to engage targets at much greater distances. Whereas the longest practical range for an air defense gun is roughly 8–10 km, some missiles can engage an incoming target tens of kilometers away. Missiles, however, have the

---

[68] "Phalanx Close-In Weapon System," video, *Military.com*, posted by "GunFun," June 15, 2012.

disadvantages of usually being much larger systems compared to guns, and their cost is far higher than guns. Missile systems are also generally less mobile compared to guns, and are more difficult to transport in strategic lift assets.

One of the best-known defensive missile systems today is the Israeli Iron Dome. Developed by the Israeli company Rafael Advanced Defense Systems, the system saw its first combat use in April 2011 when an Iron Dome battery intercepted a 122-mm rocket fired from Gaza toward southern Israel. By December 2012 the system had apparently intercepted roughly 400 short-range rockets. Deployed in battery-sized units, each costing roughly $50 million, individual Iron Dome interceptor missiles have a price of approximately $50,000. The United States and India are considering adopting the system.

The current Iron Dome interceptors are capable of engaging incoming rockets fired from up to 70 km away. This gives the system good capability against older 122-mm Katushyas and similar weapons. The system also can defend an area roughly 150-km square.[69]

Iron Dome was designed with Israel's particular defensive needs in mind. Therefore, it is a generally immobile system that defends urban areas. It is also a heavy system (since air deployment is not a consideration for the Israeli armed forces), and would be hard to transport by air in its current form. Importantly, the current version of Iron Dome is not designed to engage larger, higher speed rockets such as Smerch and WS-2, although Israel is trying to improve the system's performance against higher speed threats.

Figure 6.6. Israeli Iron Dome Firing Unit

SOURCE: Publicity photo from Israel Defense Forces.

---

[69] Revital Levy-Stein and Gili Cohen, "Iron Dome Battery Successfully Intercepts Target," Haaretz, August 13, 2008.

Another smaller and more mobile foreign defensive missile system is the Russian Pantsir-S1 system (Figure 6.7), known to NATO as the SA-22 Greyhound. This truck-mounted combination gun-missile system is intended to provide highly mobile point defenses against aircraft, UASs, air-to-ground munitions, and cruise missiles. The system is capable of engaging targets up to 20 km from the launcher, at altitudes of more than 40,000 feet. Its mobility is a desirable feature and its relatively small size means that it can be transported in large cargo aircraft.

Figure 6.7. Russian Pantsir-S1

SOURCE: "Pantsir-S1" shared via Flickr by Dmitry Terekhov, CC BY-SA 2.0.

A number of Middle Eastern countries have either acquired the system or plan to, and Brazil is considering ordering the weapon. While far more mobile than the Israeli Iron Dome, the Pantsir-S1 would not be able to engage long-range artillery rockets. Against UAS-type targets and subsonic cruise missiles, Pantsir-S1 is a formidable weapon.[70]

Although not a foreign system, we elected to include the U.S. Navy's Rolling Airframe Missile (RAM) as another example of a currently available defensive missile system (see Figure 6.8). In service since the early 1990s, the RIM-116 RAM is a short-range ship-mounted weapon intended for point defense of Navy warships. It is mounted on aircraft carriers and large amphibious ships. Designed to engage very-high-speed, sea-skimming antiship cruise missiles, the RAM is linked to the target acquisition radar of the Navy's CIWS (Phalanx) gun defensive system. As with the CIWS gun, the RAM is intended to be a very-rapid-response system that can acquire and engage high-speed incoming threats.

---

[70] "Pantsyr S1 Close Range Air Defence System, Russia," *Army-Technology.com*, undated.

Like the CIWS gun, the RAM penetrates very little into the hull or superstructure of the host ship, making it easy to mount aboard larger vessels, such as *amphibs*. The entire mount weighs roughly 12,500 pounds, and costs less than $450,000. The range is roughly 9 km against supersonic targets, using a very-high-speed (Mach 2+) interceptor.[71]

Figure 6.8. U.S. Navy Rolling Airframe Missile

SOURCE: Photo by the U.S. Navy.

As mentioned earlier, during the fighting in Iraq and Afghanistan Navy CIWS gun systems were mounted ashore on top of concrete pedestals to provide defense against incoming mortars and rockets. It may be possible to develop a similar ground mount for RAM, especially for fixed facilities such as air bases, ports, and command centers.

Laser Defensive Systems

A third possible technology solution for defense against rockets, artillery, cruise missiles, and UASs are speed-of-light weapons: lasers and directed energy. Much work has taken place in this field in the last 20 years, and as the technology has improved, the possibilities of a viable military laser defensive system grows. The United States is not alone in this field, as other countries are also experimenting and designing laser weapons, primarily for defensive roles.

An important consideration for the Army is the ability to transport and operate laser systems in a battlefield environment. This includes providing adequate power sources and cooling systems to cope with the excess heat that lasers generate. One reason that the world's better navies are ahead of armies in the development and fielding of laser systems is the fact that ships have enormous

---

[71] "RIM-116 Rolling Airframe Missile," United States Navy Fact File, *Navy.mil*, last updated November 19, 2013.

power-generating capabilities. Additionally, by definition ships provide the required transport to move their defensive lasers with them.[72]

A potentially significant advantage of laser defensive systems over guns and defensive missiles is their *multishot* ability. If the laser works, if it has sufficient power, and if the atmospheric conditions permit, a laser defensive system can take literally tens of shots before its power system would need refueling or the system's cooling given a pause to bleed off excess heat.

This section will not discuss the technological pros and cons of laser defensive systems. Rather, what follows is an example of the types of weapons that are being developed in other countries. It should be noted that compared with guns and defensive missiles, less foreign development in land-based laser defensive systems is under way. In addition to the German example that follows, the Chinese and Russians are conducting some development of laser weapons, but the unclassified nature of this report does not allow those examples to be included in this document.

The German Rheinmetall High-Energy Laser is a 50-kilowatt (kW) laser that has successfully engaged aerial targets during tests in 2011 (see Figure 6.9). The system is being specifically designed to perform a counter–UAS, artillery, mortar, and rocket mission.[73]

Although initially intended for defense of fixed points, the size and weight of the system makes it feasible to incorporate it into a large military truck, perhaps with a vehicle following to provide additional power. The Germans have stated that the test results were so favorable that a larger, more powerful 100-kW system could be developed in the next few years.

Figure 6.9. Rheinmetall 50-kW Defensive Laser System

SOURCE: Publicity photo from Rheinmetall.

---

[72] Information on U.S. laser systems was provided by various industry representatives at a meeting at RAND's Santa Monica, California, office in March 2013.

[73] "Rheinmetall Demonstrates 50kw HEL Laser," *Optics.org*, December 19, 2012.

In testing thus far, the Rheinmetall system has engaged airborne targets at distances of 10 km or fewer. Depending on the power of the final system and the prevailing atmospheric conditions during a specific engagement, that distance could be lengthened in the future.

Insights and Implications for the U.S. Army

This chapter focused on defensive systems to counter the growing threat from UASs, cruise missiles, and long-range artillery rockets. As can be seen from the examples shown in this chapter, a number of other countries are also developing defensive systems that range from guns to missiles to laser systems, and it should be noted that this chapter is a representative sampling of what is happening in other parts of the world.

The Army is studying the need for improved protection against rockets and missiles. Guns, missiles, and lasers are being considered for addition to the suite of defensive systems currently available.

There are advantages and disadvantages associated with each of the defensive technologies highlighted in this chapter. Guns are relatively cheap and fairly easy to deploy, but they have a short range and cannot cope with very-high-speed targets. Defensive missiles are more expensive and tend to be larger and heavier than guns, but they have greater range and have a better chance of defending against the current generation of long-range rockets that approach at speeds of Mach 5 or higher—although not all defensive missile systems (e.g., the original version of Iron Dome) can deal with that high-end threat. Lasers have considerable potential, but they are not panacea weapons. Lasers are degraded by rain and other atmospheric conditions, and there have been false starts in the past with supposedly promising laser defensive systems, such as the canceled U.S.-Israeli Tactical High Energy Laser (THEL) that was terminated in 2005 following more than $300 million in research and development funds spent on the program up to that point.

Perhaps the most appropriate insight for the U.S. Army at this point of time is that there are a number of defensive technologies available, each with advantages and disadvantages, including some promising foreign systems. As the Army searches for possible counters to the growing long-range fires threat, it should continue to explore its options until and unless a clear standout system appears. In addition to the kinetic systems covered in this chapter (guns, missiles, and lasers), the Army should also consider passive measure, such as decoys, to limit the effectiveness of hostile rockets, missiles, and UAVs, as well as the possibility of using electronic attacks on the onboard guidance systems (if any) of this class of threats.

7. Infantry Squads

In 2011, the commanding general of the U.S. Army Maneuver Center of Excellence, Major General Robert Brown, recognized the dismounted squad as being the "foundation of the decisive force." Yet this is the squad level in which the U.S. Army experiences the least amount of decisive advantages with respect to current and foreseeable threats.[74]

The U.S. Army is not alone in its desire to achieve overmatch within the fundamental building blocks of a ground force. Lessons from recent conflicts, advancements in technology, and shifting geopolitical priorities have led armies around the world to invest in researching and designing soldier and squad modernization programs. Some of these programs appear to primarily focus on technological advances to improve a squad's current capabilities and capacity. Other programs highlight shifting priorities for a squad's function, while still others include associated changes to organization, training, vehicles, and employment principles.

This chapter analyzes a selection of infantry squad and soldier modernization programs from around the world in order to support the Army with its equipment modernization investment decisions. While the primary objective of this chapter is to highlight other equipping solutions for service-dismounted squads, the employment of that equipment and decisive overmatch is intimately tied to a squad's task organization and, at times, the effects of its transportation options. Thus, two sections are dedicated to a short analysis of manning and IFV preferences. Another section briefly compares squad-level counterdefilade weapon systems. The chapter concludes with a review of the common themes across all reviewed services, a collection of notable features that a model squad system may contain, and some equipping-related considerations for Army decisionmakers.

As was the case in the previous chapters, there are differing design philosophies among the world's armies regarding how to organize and equip infantry squads. Even within the U.S. military, the Army and Marine Corps take different approaches to the size of a squad. As the Army considers future squad organization and equipment, it should carefully consider the missions that it expects squads to perform.

Rifle Squad Organizations

Since the end of World War II, the U.S. Army rifle squad has undergone at least a half-dozen organizational transformations. From a 12-man squad organized into sections for leadership (leader and assistant leader), scouts, fire, and maneuver (displayed in Figure 7.1), the Army squad has been adjusted based on findings from various studies. It was reduced to 9, increased to 11, and

---

[74] Rob McIlvaine, "Squad Needs 'Overmatch' Capability," U.S. Army, October 13, 2011.

brought back to 9, along with various organizational arrangements throughout the years.[75] The following examines how select armies around the world currently organize and man their dismounted rifle squads.

Figure 7.1. World War II U.S. Army Rifle Squad

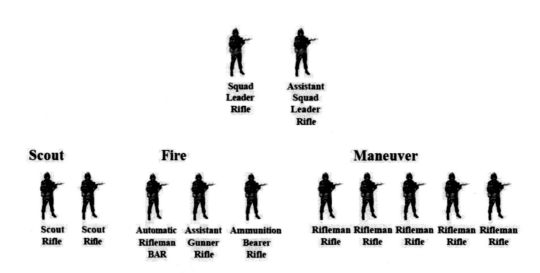

SOURCE: Reprinted from Karcher, 1989.

A comparison of seven services (including the U.S. Army and Marine Corps) reveals that none appears to man and organize squads the same way. As described in Table 7.1, squads (referred to as *sections* by some armies) and fire teams (sometimes classified as *groups*) are composed differently around the world.

---

[75] Timothy Karcher, "Enhancing Combat Effectiveness, the Evolution of the United States Army Infantry Rifle Squad Since the End of World War II," master's thesis, United States Army Command and General Staff College at Fort Leavenworth, Kan., 2002.

Table 7.1. Dismounted Rifle Squad Manning and Organization of Select Services

	Fire Team (Group) 1					Fire Team (Group) 2				Fire Team (Group) 3			
USA	Sqd Ldr	Team Ldr	Rifleman	M203 Grenadier	LMG Gunner	Team Ldr	Rifleman	M1023 Grenadier	LMG Gunner				
USMC	Sqd Ldr	Team Ldr	Rifleman	M203 Grenadier	Auto Rifleman	Team Ldr	Rifleman	M1023 Grenadier	Auto Rifleman	Team Ldr	Rifleman	M1023 Grenadier	Auto Rifleman
UK		Section Cmdr	LMG Gunner	M203 Grenadier	LMG Gunner	Section 21C	LMG Gunner	M203 Grenadier	LMG Gunner				
CAN		Section Cmdr	Rifleman	M1023 Grenadier	LMG Gunner / Rifleman	Section 21C	Rifleman	M203 Grenadier	LMG Gunner / Rifleman				
AUS		Section Cmdr	Scout	M203 Grenadier	LMG Gunner	Group Leader	Scout	M203 Grenadier	LMG Gunner				
FRA	Sqd Ldr	Team Ldr	Sniper (7.62)	51-mm Grenadier	LMG Gunner	Team Ldr	Foreman (AT4CS)	Rifleman (AT4CS)		Gunner	Driver		
GER		Section Cmdr	Rifleman	Grenadier	LMG Gunner	Team Ldr	Rifleman	Grenadier	LMG Gunner	Gunner	Driver		

NOTE: This figure was created using a combination of sources, including information provided by foreign military liaisons at the U.S. Army Training and Doctrine Command Headquarters. This table represents a baseline of task-organizing rifle squads from which numerous permutations may occur; some of these are discussed conceptually throughout this chapter.

Squad Leadership

One of the most obvious differences in Table 7.1 is that some armies (UK, Canada, Australia, and Germany) appear to place the squad/section leader into a particular fire team/group. A similar way to look at it is to designate one of the team leaders as being the squad leader. This suggests that the squad leader is tasked with directing the individuals within a single fire team along with another fire team leader. As such, it may be advantageous to ensure that the squad leader is external to the fire teams in order to focus efforts on directing individual team leaders and maintain a higher-level situational awareness. However, one of the consequences of structuring the squad with a leader embedded into a fire team is that in a two-team squad, an assistant squad leader, or 2IC, is essentially designated by default. Squads without this structure (the U.S. Army, U.S. Marine Corps, and France) do not have an organizationally defined succession of command for when the squad leader is removed from action. Similarly, some tactical missions, particularly in patrolling and distributed operations, benefit from having assistant patrol leaders assume responsibilities beyond those of an average team leader. In order to realize the advantages of both structures, there may be value in considering an organization that includes both a squad leader and assistant squad leader who are both separated from the subordinate fire teams. This leadership solution would look similar to the organization of the American Army squad's senior members during World War II.

Squad Size

Of the selected services reviewed, the size of the U.S. Army rifle squad appears to be about average at nine. Four of the seven squads are composed of only eight men, and the only services with larger squads are Canada (10) and the U.S. Marine Corps (13). One of the common factors discussed along with squad size, to include one of the primary reasons for decreasing the squad size from 12 after World War II, is control.[76] The creation and improved training of fire team leaders should eliminate the need for a squad leader to control each individual member of the squad instead of only the respective team leaders. The USMC's 13-man squad is broken into three separate fire teams based on the assumption that an individual is capable of tactically controlling up to three distinct groups at a time.[77] Further, all of the soldier and squad modernization programs being pursued by a number of countries described in subsequent sections unambiguously attempt to improve command and control at the squad level.

While the limitations of control and accountability suggest the need to keep squad size relatively small, there are numerous advantages to the larger-sized squads, particularly ones with three fire teams. First, an individual squad's resiliency as a system increases as the number of personnel increase. Depending on the environment and the weight of gear, an incapacitating injury may require two or three healthy individuals to move, protect, and begin treating a single casualty. Thus, one hurt individual could effectively reduce an eight- or nine-man rifle squad to one fully capable fire team—in effect removing the squad as a system from the fight altogether. The three teams in the USMC's rifle squad may absorb casualties better while remaining in the fight as a squad. Relatedly, illnesses, training injuries, and a number of other factors are likely to keep many squads from deploying and operating at full strength. A larger squad is more capable of maintaining a baseline capability when losses are suffered.

Another advantage of having three fire teams is that the structure lends itself to fire and maneuver techniques by providing greater firepower from two fire teams, which allows the third

---

[76] Karcher, 2002.

[77] The USMC rule of threes is: three Marines led by a fire team leader, three fire teams in a rifle squad, three squads in a rifle platoon, three rifle platoons in a rifle company, and three rifle companies in an infantry battalion.

to maneuver. Three fire teams are also conducive to task organizing the squad into assault, support, and security elements, a useful technique during assaults in urban terrain and some patrolling operations. The two-fire-team structure with fewer soldiers may be more limited in conducting these types of tactical squad tasks unilaterally.

The review of soldier and squad modernization programs that follows highlights the constant concern regarding equipment weight. Increased weight carried by each soldier decreases the squad's mobility, requires additional sustainment consumables (food and water), and risks excessive injuries, making a squad less lethal and overall less effective. In addition to individual equipment, a squad is likely to carry a number of additional squad items. Additional items may include unmanned aerial vehicles (UAVs) and UGVs, communications equipment, signaling devices, additional weapons and ammunition, cameras, breaching tools, and force protection items, such as Remote Control–Improvised Explosive Device (RC-IED) jammers and mine detectors. These items, and many more, may be considered requisite for a squad's mission regardless of the size of the squad. Larger squads allow the additional equipment to be spread among more soldiers, thereby reducing the individual load of each person. Table 7.2 demonstrates how a rifle squad can easily carry more than 100 pounds of squad equipment and the resulting 33 percent decrease in additional load per soldier with a third four-man fire team. Each column excludes the squad leader. In this example the average weight that soldiers in a squad with two fire teams will carry is an additional 12.75 pounds, while those in a squad with three fire teams have an additional 8.5 pounds to bear for the same equipment. This decrease in shared load to be carried becomes increasingly significant as the distance traveled and times of operations are extended. The difference also exceeds the weight of a gallon of drinking water.

Fire Team Composition

Table 7.1 shows that it is common for most infantry fire teams to include one light machine gun and one grenade launcher.[78] It is also common for fire teams within a squad to be composed of identical billets and weapons. An advantage to doing so is that common training, organization, and employment will facilitate the movement from one position within a squad to another with little additional training requirements. The one outlier listed in Table 7.1 is France. The French squad consists of one fire team with a sniper, light machine gun gunner, and grenadier and a second fire team consisting of two riflemen armed with antitank rockets. Both of which are led by separate fire team leaders. Reportedly, the German Army is considering replacing one of its fire team's light machine guns with a medium machine gun and one grenadier with a designated marksman. Breaking the symmetry of fire team composition may allow for individuals to specialize in a specific skill or function, but, as mentioned earlier, that could hinder the transition from positions within the squad. Similarly, most infantry platoons are made up of equally

---

[78] Note that Table 7.1 represents the USMC's recent replacement of light machine guns with Infantry Automatic Rifles at the fire team level.

composed squads. One notable exception to this is the Israeli infantry platoon. These platoons are composed of three distinctive squadrons, which are uniquely task organized with different positions and weapons in order to satisfy *charging*, *following*, or *cover* responsibilities.

Table 7.2. Weight Distribution of Common Squad Items Beyond Individual Equipment Loads

Squad Item	Weight	Distributed Weight per Soldier (8-Man Squad)	Distributed Weight per Soldier (12-Man Squad)
RC-IED jammer	20 lbs[a]		
Handheld metal detector	8 lbs[b]		
Collapsible litter	15 lbs[c]		
24" bolt cutters	5 lbs	12.75 lbs	8.5 lbs
Portable UAV	17 lbs[d]		
2 AT4s (15 lbs each)	30 lbs[e]		
2 Claymores (3.5 lbs each)	7 lbs[f]		
Total weight	102 lbs		

[a] John Keller, "JIEDDO Seeks to Shrink Soldier-Worn IED Detector Technology to Less Than 20 Pounds," *Military and Aerospace Electronics*, October 18, 2012b.
[b] "AN/PSS-14 Handheld Standoff Mine Detection System (HSTAMIDS)," *GlobalSecurity.org*, undated.
[c] "Litter Evacuation," Chapter Nine in *Medical Evacuation in a Theater of Operation Tactics, Techniques, and Procedures*, Field Manual 8-10-6, Washington, D.C.: Headquarters, Department of the Army, April 14, 2000.
[d] "Honeywell Unmanned Aerial Vehicle Included in Army's Contract for Brigade Combat Team Modernization Increment 1," press release, Honeywell, August 24, 2010.
[e] "M136 AT4," *Military Analysis Network*, updated January 8, 1999.
[f] *Antipersonnel Mine M18A1 and M18*, Field Manual 23-23, Headquarters of the Department of the Army, January 1966.

Infantry Fighting Vehicle Effects on Squads

U.S. Army infantry squads must be able to conduct operations in various combat environments, whether mounted or dismounted. This is a common requirement for infantry squads around the world that employ a wide range of vehicles. This section reviews some common IFVs and considers the effects of their troop-carrying capacity on the squads in which they are designed to transport. For a more comprehensive comparison of vehicles and their full capabilities, see Chapter Two.

The USMC's Expeditionary Fighting Vehicle (EFV) was never completed as planned but was designed to hold a 13-man rifle squad plus four attachments, for a total of 17.[79] Currently, the Marine Corps employs the Amphibious Assault Vehicle Personnel Model 7A1 (AAVP-7A1),

---

[79] "The USMC's Expeditionary Fighting Vehicle (EFV)," *Defense Industry Daily*, June 26, 2012.

which is designed to hold 21 combat-loaded Marines,[80] although additional equipment often makes it difficult to obtain this maximum capacity. See Figure 7.2.

Figure 7.2. Infantry Fighting Vehicles and Dismount Capacity

Service	Vehicle	Crew			Dismounts									
		Driver	Commander	Gunner										
AUS Army	ASLAV	Drv			Section Cmdr	M203 Grenadier	LMG	Section 2IC	M203 Grenadier	LMG				
CDN Army	LAV III	Drv			Section Cmdr	M203 Grenadier	LMG	Rifleman	Section 2IC	M203 Grenadier	LMG			
UK Army	F510 Warrior	Drv			Section Cmdr	M203 Grenadier	LMG	Rifleman	Section 2IC	M203 Grenadier	LMG			
AUS Army	M113	Drv			Section Cmdr	M203 Grenadier	LMG	Rifleman	Section 2IC	M203 Grenadier	LMG	Rifleman		
U.S. Army	M2 Bradley	Drv	Commander	Gunner	Squad Leader	Team Leader	M203 Grenadier	LMG	Rifleman	Team Leader	M203 Grenadier	LMG		
U.S. Army	M1126 Stryker	Drv	Commander		Squad Leader	Team Leader	M203 Grenadier	LMG	Rifleman	Team Leader	M203 Grenadier	LMG	Rifleman	
UK Army	FV432	Drv	Commander		Section Cmdr	M203 Grenadier	LMG	Rifleman	LMG (-)	Section 2IC	M203 Grenadier	LMG	Rifleman	LMG (-)
U.S. Marine Corps	AAVP7	Drv	Commander	Gunner	Squad Leader	Team Leader	M203 Grenadier	IAR Gunner	Rifleman	Team Leader	M203 Grenadier	IAR Gunner	Rifleman	Team Leader	M203 Grenadier
					IAR Gunner	Rifleman	Att 1	Att 2	Att 3	Att 4	Att 5	Att 6	Att 7	Att 8	Notional Examples

SOURCE: Data in table from V. Sattler, and M. O'Leary, "Organizing Modern Infantry: An Analysis of Section Fighting Power," *The Canadian Army Journal* Volume 13, No. 3, 2010. Data about the USMC AAVP7 from "AAV-7," *Marines.com*, undated.

After working with data, it became apparent how the troop capacity of various vehicles impacts the ability of a squad to operate as a system in the manner it was designed. Table 7.3 assesses 11 different vehicles and their dismount capacity impact on the squad a vehicle is intended to support. A negative value in the column titled "Impact" represents the number of soldiers a squad embarks the vehicle without. A positive value represents the available space for additional enabling personnel to attach and deploy with the squad.

An issue with the Bradley IFV is the fact that the vehicle cannot accommodate nine dismounting infantrymen, assuming that the crew remains in the vehicle when the rest of the

---

[80] "AAV-7," *Marines.com*, undated.

squad dismounts. Conversely, the Stryker IFV can accommodate the vehicle crew and a complete nine-man squad that dismounts.

Table 7.3. Impacts of IFV Capacities on Squad Employment

Service/Country	Vehicle	Capacity	Squad Size	Impact
U.S. Army	M2 Bradley	6	9	–3
Canada	LAVIII[a]	7	10	–3
Australia	ASLAV[b]	6	8	–2
United Kingdom	FV510 Warrior	7	8	–1
U.S. Army	M1126 Stryker	9	9	0
France	AMX-10P	8	8	0
Germany	ARTEC Boxer	8	8	0
United Kingdom	FV432[c]	10	8	+2
Australia	M113[d]	10	8	+2
Germany	Rheinmetall Landsysteme Condor[e]	12	8	+4
U.S. Marine Corps	AAVP7	21	13	+8

[a] Light Armoured Vehicle (LAV) III," Canadian Army, modified October 23, 2013
[b] "Australian Light Armoured Vehicle," Australian Army, undated.
[c] "FV432 APC," *Military Factory*, last update July 7, 2013.
[d] "M113 AS4 Armoured Personnel Carrier," Australian Army, undated.
[e] "Rheinmetall Landsysteme Condor," *Military Factory*, last updated June 24, 2013.

The impact column of Table 7.3 suggests that an IFV/APC capacity of less than squad size negatively affects the ability of a squad to operate as a system upon dismounting the vehicle. This does not, however, take into account how the vehicles and platoon operate as a larger system in which the division of squads into separate vehicles may be intended. Even so, it is not unreasonable to expect that most systems, whenever possible, should be moved as a complete unit. It would not make sense, for example, to move most weapon systems into combat while disassembled and transported in separate vehicles only to be combined later at the moment employment is required.

It should be noted that additional squad- and soldier-equipping initiatives have continued to increase both the weight and volume of weapons and gear that soldiers haul along with them when moving in a medium-threat environment. Most of the vehicles listed in Table 7.3, for example, were designed long before squads were ever projected to operate with tactical UAVs, RC-IED jammers, and handheld metal detectors. Numerous robotic devices are also currently being developed and tested for future support to infantry squads (see Chapter Eight). While these devices are likely to enhance a squad's capabilities when dismounted, they will add additional transportation requirements, compromising available transport capacity even further.

Squad and Soldier Modernization Programs

Program Selection Criteria

Squad and soldier modernization programs are currently under way around the world. We identified at least 30 separate programs that exist in differing stages of development and employment in nearly as many countries. Some of the programs do not have widely available information. Others exist in such nascent stages that specific systems and devices have not yet been developed and tested and may remain as vague capability objectives. Other programs are introducing technology and gear that will advance a foreign service's infantry squad to a comparable level of equipping as current U.S. Army squads. These programs were left out of the following analysis. This section includes an examination of 11 separate modernization programs that are believed to include innovative uses of different technologies, provide significant battlefield advantages, or empower squads to go beyond simply improving current capabilities, possibly enabling the development of new tactics and techniques.

Before the providing information on individual countries, we highlight the issue of the weight that an individual infantryman is expected to carry today. In interviews with the liaison officers from France, Israel, the UK, Canada, Germany, and Australia conducted at U.S. Army Training and Doctrine Command (TRADOC) Headquarters, all of the representatives noted that their armies are concerned with the weight of the equipment that today's dismounted infantrymen are expected to carry. While recognizing that a lot of innovative and useful weapons, body armor, and other equipment have been fielded in the past decade, they all expressed concern that today's infantryman is carrying too much weight. Several stated that their armies are now in the process of making a concerted effort to reduce the burden on the infantryman

France

One of the more advanced modernization programs is the French system titled FÉLIN (Fantassin à Équipements et Liaisons Intégrés). The French Army set a target weight of 25 kg (55 pounds) for the overall system, which includes an individual soldier's weapon, ammunition, and enough food and water for 24 hours. New equipment includes portable computers, radios, clothing, and advanced helmets. Multiple versions of the system were intended to be available for different levels of command, but every soldier is provided with a radio and GPS device. The helmet includes a monocular with two LED (light-emitting diode) displays (each three square centimeters), a light-intensifying camera, and an OH-295 osteo-microphone built into the headband, which uses vibrations from the wearer's skull.

The FÉLIN weapon systems include the Giat FAMAS F1 5.56-mm assault rifle, the Giat FR-F2 7.62-mm sniper rifle, and the FN Herstal Minimi 5.56-mm light machine gun (see Figure 7.3). Weapon sights are linked to the communications system and can send real-time video to the soldier, allowing him to aim around corners without exposing his body, and to other members of the squad, enhancing the ability to communicate and share information. The system's binoculars include a laser rangefinder, a digital magnetic compass, and an uncooled thermal imaging channel that reduces weight.[81]

Figure 7.3. FÉLIN-Equipped Soldier Aims Weapon

SOURCE: Photo by Daniel Steger, CC BY-SA 1.0.

Germany

Ensuring that every soldier has access to real-time information was also a fundamental principle in the development of the German Infanterist der Zukunft–Erweitertes System (IdZ-ES); see Figure 7.4. Among the numerous advancements, every infantryman is equipped with a GPS, electronic compass, inertial navigation system,[82] centralized and computer-managed power source, and improved ballistic protection that weighs less than prior versions. Modified clothing seeks to protect soldiers from extreme temperatures, nuclear and biological threats, and insects while simultaneously reducing the thermal signature of troops so that they are less easily identified using thermal imaging devices.[83] Squad commanders also have access to a portable computer tablet attached at the waist and a Toughbook laptop for additional planning support that is connected

---

[81] "FELIN (Fantassin à Équipements et Liaisons Intégrés)—Future Infantry Soldier System, France," *Army-Technology.com*, undated.

[82] "Rheinmetall Starts Gladius Soldier System Deliveries to German Army," *Army-Technology.com*, March 14, 2013.

[83] Paolo Valpolini, "The Weight of Intelligence," *Armada International*, April 1, 2010.

through ultra-high-frequency radio to the rest of the squad. Soldiers in the squad can access that information in the form of voice, data, and image either through a helmet-borne eyepiece or a "dogbone" handheld display, which are both connected via cable to a central processor worn on a soldier's back. The cable connection to the helmet device removes the need for an additional power source to be worn on the helmet itself.[84]

Weapon sights, however, are linked to the central processor with the use of Bluetooth wireless technology.[85] Similar to the FÉLIN system, around-the-corner observation and firing of the G36 rifle is made possible through the use of the Aimpoint Concealed Engagement Unit.

Functions available from buttons on the weapon itself include "push to talk" along with a laser rangefinder controlled with buttons near the trigger guard.[86] Additional functions that the German Army had previously shown interest in including in the IdZ-ES systems include RC-IED jammers and acoustic shot detection technologies.[87]

Figure 7.4. Infanterist der Zukunft

SOURCE: Photo shared via Flickr by Bundeswehr/Rott, CC BY-NC-ND 2.0.

United Kingdom

The UK's Future Integrated Soldier Technology (FIST) is not expected to begin entering the service until between 2015 and 2020, but it is projected to share many of the same features as the French and German systems. These similarities include a degree of data and voice interconnectedness that allows images from other members of the squad, higher headquarters, and

---

[84] "A Soldier's Burden," *Jane's Defence Weekly*, last posted March 7, 2013.

[85] "A Soldier's Burden," 2013.

[86] Valpolini, 2010.

[87] "A Soldier's Burden," 2013.

intelligence collection platforms such as drones to be shared, potentially through a range of devices, such as helmet-borne monoculars, wrist-worn devices, handheld displays, and portable computers. Linking weapon sights to such displays would also allow for firing the SA80 assault rifle from protected positions, minimizing exposure. Another similarity to the German system is that the UK will attempt to provide improved protection and reduced visual, radar, and infrared signatures through advanced clothing that may use built-in wires to help connect the subcomponents of the FIST system if Bluetooth technology is not applied.

One of the major differences between the proposed UK system and the French and German versions is that in the case of the British, it is not expected that every infantry soldier would be fitted with a FIST system. Rather, the unit commander may modify which soldiers are given the system to accommodate a particular mission and situation. Another intention is to reduce the amount of weight carried by the average soldier through the use of noncooled observation and weapon sights. Additional force protection considerations include a warning device for any readings of nuclear, biological, or chemical (NBC) threats, which would also send NBC reports to national-level command centers.[88]

Israel

The Israeli military has contracted its soldier and squad modernization program to Elbit Systems, which has developed the Dominator Integrated Infantry Combat System (IICS). Similar to the previous systems, Elbit advertises that the Dominator system will allow soldiers to view a real-time common operating picture (COP) on personal displays, send and receive data and images, and enhance operational planning and debriefing through technological solutions. At the center of the system is the Personal Digital Unit, which allows for data processing as well as storage and is linked to all other elements of the Dominator system. Images and data can be accessed by individuals through a helmet, a handheld eyepiece, or an eight-inch planning display.

Included in the Dominator system are the VIPeR stair-climbing unmanned ground vehicle (see Figure 7.5) and the Skylark mini UAV. Intended to be portable and configurable for a variety of uses, these devices are to be carried by the squad to enhance its ability to provide its own reconnaissance and information gathering. Another key characteristic is that all of the components are designed to be modular, so that equipping solutions can be tailored for specific billets and mission requirements.[89] Regardless of how the system is set up, individual location reporting embedded in the system could inform the COP, and it is powered from a central battery that is said to support the system for 24 hours.[90]

---

[88] "FIST—Future Infantry Soldier Technology, United Kingdom," *Army-Technology.com*, undated.

[89] *DOMINATOR®, Enhanced Combat Effectiveness for Infantry Units and Special Forces*, St. Netanya, Israel: Elbit Systems, 2009.

[90] Tamir Eshel, "Israeli Multi-Purpose Tank Ammo Redesigned to Fit the 125mm Gun of the T- 90S," *Defense Update*, March 27, 2012a.

One subcomponent that was reportedly being considered for addition to the Dominator system was the Soldier Navigation (S-Nav) system also created by Elbit. The S-Nav would detect altitude changes and measure a soldier's movement through a variety of sensors that would ensure that the soldier's position is always known even when GPS signals are not available. Suggested to be about the size of a cell phone, this device would greatly improve navigation and position reporting while in buildings, subterranean areas, and thick jungle or forest environments.[91]

Figure 7.5. Israeli VIPeR Unmanned Ground Vehicle

SOURCE: Publicity photo from Elbit Systems.

NOTE: The Israeli VIPeR unmanned ground vehicle carries a variety of sensors and payloads for the Israeli infantry squad.

Norway

In Norway the Norwegian Modular Arctic Network Soldier (NORMANS) system has tested positively since 2010. NORMANS comes in two versions, the Light, which is issued to every soldier, and the Advanced, which is used by small-unit leaders. In the Light system, each soldier is given a cell-phone-sized visual display with a wrist-worn controller that provides individuals with a map of their positions along with those of the other members of their units and navigational aids for predetermined routes planned by the squad. In the Advanced system, small-unit leaders are linked via very high frequency to the next higher-level battle management system. The application of these linked visual planning displays was demonstrated during testing in 2010, when small-unit leaders developed a scheme of maneuver and digitally sent the overlay to each member of the unit; consequently, all members were able to move to assigned positions without having to physically come together to be briefed about the new plan, and leaders could monitor precisely where each soldier was physically located without tangibly seeing each person. It was also stated that during testing, the personal display's ability to download and view aerial photos and maps assisted in selecting routes that provided the best cover and concealment.[92]

---

[91] Valpolini, 2010.

[92] NORMANS Trials Update," *Soldier Modernisation*, Vol. 5, May 2010.

The NORMANS system was designed to network not only with other radio, sensor, and vehicle systems but also with Norwegian simulation systems that may be valuable in training. Given the need to minimize weight carried by individual soldiers, Norwegians sought to reduce requirements primarily within clothing, protection, and weapon systems. For example, Norway has purchased the 5.56-mm HK416 to replace the previously used 7.62-mm weapon, reducing the size and weight required by soldiers, who are now carrying smaller and lighter ammunition.[93]

Poland

The Poland soldier and squad modernization program dates back to 2006, when the Indywidualny System Walki (ISW) Tytan was begun (see Figure 7.6). One of the key features of the ISW Tytan program was that it was to be easily modified and upgradable so that new technologies could be quickly added or used to replace subcomponents within the system in the future. Three basic versions were designed to accommodate dismounted infantrymen, mechanized soldiers, and reconnaissance and specialized units.[94] Advances to sights and C4I systems are similar to those of systems described previously, as Poland sought close coordination with other military and contractors that had designed systems, such as FÉLIN.[95]

Figure 7.6. Polish Soldiers Display Improved Sights and Communications Equipment as a Part of the ISW Tytan System

SOURCE: Publicity photo from Polish Defence Holding.

---

[93] "Soldier Systems Modernization in the Norwegian Armed Forces," *Soldier Modernisation*, Vol. 7, June 2011.

[94] "Tytan at a Turning Point," *Soldier Modernisation*, Vol. 3, June 2009.

[95] Valpolini, 2010.

Spain

Also initiated in 2006, the Spanish Combatiente Futuro (COMFUT) program incorporates in different ways a number of the advancements described earlier (see Figure 7.7). For example, the COMFUT system also includes embedded GPS and digital magnetic compasses, but it combines the data for more than navigation assistance. The system provides soldiers with an acoustic and visual alarm when the compass detects that a weapon is pointed toward another member of the unit, which is intended to prevent fratricide. The weapon also sends the soldier information regarding how many rounds remain in a magazine.

The COMFUT advanced weapon sight has an internal battery that provides four to five hours of continuous operation but can also be connected with a soldier's main battery, worn on the body. The separate sight system uses Bluetooth technology to send data regarding the digital compass and fire control system, while another device sends video imagery. Many of the other C4I improvements are similar to the German IdZ-ES system.

Figure 7.7. Spanish Soldier Using Advanced Sights from the COMFUT System

SOURCE: Publicity photo from Ministerio de Defensa de España, at mde.es.

Italy

The Italian Soldato Futuro program shares many of the same C4I improvement aims as the previously discussed systems (see Figure 7.8). One of the features that help it accomplish this is the use of advanced networking services, such as Mobile Ad Hoc Network (MANET) and Multi-Net voice operations. A mobile wireless network can be established with up to 50 different units with a data capability that facilitates the transmission of streaming video.[96] Monocular displays

---

[96] Paolo Valpolini, "Italy's Soldato Futuro Moves Towards Production," *Soldier Modernisation*, Vol. 8, January 2012.

allow soldiers to view low-light-level images, maps, and other digital messages sent from other soldiers or unit leaders. In 2010, improved designs were intended to reduce the cables required to link all of the associated components in all three existing versions, including different arrangements for commanders, grenadiers, and riflemen.[97]

Figure 7.8. Italian Soldato Futuro System

SOURCE: Publicity from Selex ES.

South Korea

The South Korean Future Warrior program appears to have two primary objectives: improving squad-level C4I through advanced networking and optimizing NBC detection and protection. Modularity has also been reported to be an important characteristic of soldier and squad systems. One addition to the networked computer displays not advertised in other systems is that South Korean squad leaders may be able to access and manipulate digital information through the use of voice recognition software along with controllers similar to those described in other soldier modernization programs. Bone-conduction earphones and neck microphones are additional features to the advanced helmet, which also has a digital head-mounted display, night vision monocular, and gas mask.

An important sustainment feature being pursued is the integration of fuel cell and lithium-ion batteries to create a hybrid fuel cell along with a battery management system. To enhance personal

---

[97] Valpolini, 2010.

protection, sensors monitor biological and environmental conditions, such as body heat, acceleration, electrocardio information, and surrounding temperature and noxious gasses.[98]

Russia

While detailed information regarding Russia's soldier and squad modernization program may not be available until the program is demonstrated at the Russia Arms Expo in September 2013, some program testing and objectives suggest that it has much in common with other international programs. After initially considering the purchase of France's FÉLIN system in 2011, the Russian Army decided to develop its own soldier system named Ratnik, which is said to include more than 40 new components, including communications equipment, power supplies, weapons, and sights (see Figure 7.9).

Figure 7.9. The Russian Ratnik System

SOURCE: Photo by Vitaly V. Kuzmin, CC BY-SA 3.0.

United States Marine Corps

The U.S. Marine Corps seems to have taken a slightly different approach to soldier modernization compared with many of the programs described in this section. In 2008 the program manager for the Marine Expeditionary Rifle Squad (MERS) program stated: "The best computer in the Marine rifle squad is 13 thinking, educated, trained Marines capable of rapid decision making in any geographical area." While the MERS program does include equipping solutions, including advanced communications and computer systems, the primary focus for improving the squad as a

---

[98] "Korea's Future Warrior," *Soldier Modernisation*, Vol. 7, June 2011.

system seems to be through the integration of squad gear, such as human performance, equipment limitations, and small-unit leader training.[99]

In 2007 a facility for squad equipment integration testing was established and named Gruntworks.[100] Gruntworks is focused on the integration of equipment and the Marines who will wear, carry, and employ new systems. This lab tests how everything from armor to batteries fit in the Marine rifle squad and how those items may affect the squad's performance or ability to fit into different vehicles.[101] The Gruntworks facility ensures that equipment for the rifle squad is kept "simple, reliable, and trainable."[102]

Another noteworthy objective from the Marine Corps Warfighting Lab has been the "Lighten the Load" initiative to reduce weight required to be carried by infantrymen. In addition to advanced batteries and more-efficient communications systems being pursued by most services, the Marine Corps is seeking to minimize other consumables, namely water (see Figure 7.10). Being both one of the most important and heaviest items carried by infantry squads, water creates increasingly challenging logistical problems as the time and intensity of operations increase. The Marine Corps is seeking to reduce the need to carry significant amounts of water by improved foraging techniques and water purification systems, which are to be applied at the squad level.[103]

Figure 7.10. Reducing Weight in Marine Rifle Squads

SOURCE: U.S. Army photo by Sergeant Michael J. MacLeod.

NOTE: Marine rifle squads may reduce weight by carrying less water and relying on irrigation ditches, streams, and lightweight water purification systems.

---

[99] "Moving Forwards," *Soldier Modernisation*, Summer 2008.

[100] Valpolini, 2010.

[101] Monique Randolph, "New Digs Equal New Opportunities for Gruntworks Team," Marine Corps Systems Command, *MarCorSysCom.Marines.mil*, March 19, 2013.

[102] Randolph, 2013.

[103] J. R. Wilson, "Sea Soldier's Load," *DefenseMediaNetwork.com*, November 9, 2010.

Counterdefilade Weapon Systems

There are a number of foreign systems that exist or are in development, but most seem to fall into one of four categories described briefly below by taking a closer look at specific systems.

Daewoo K11

Most similar to the XM25 is the Daewoo K11, which includes both a 5.56-mm rifle and a 20-mm grenade launcher (see Figure 7.11). The dual-caliber air-burst weapon allows for the 20-mm grenade fuses to be preprogramed through an electronic fire control unit. With a magazine capacity of five 20-mm rounds and a standard M16-type rifle built into the same system, a soldier transitions from weapons through a selector switch and fires both systems with the same trigger assembly.

The sight unit includes a laser rangefinder, ballistic computer, and day and infrared sighting channels. Similar to the XM25 concept, 20-mm grenades use the data from the fire control unit to explode above or next to targets.[104] In 2010 the United Arab Emirates was believed to be the first country to order exports of the K11 from South Korea.[105]

Figure 7.11. South Korean K11 Dual-Caliber Air-Burst Weapon

SOURCE: Photo by Cinnamontrees, CC BY-SA 3.0 Unported.

Multi-Purpose Rifle System

The Israeli Multi-Purpose Rifle System (MPRS) provides a different approach to engaging targets behind cover and inside structures (see Figure 7.12). Instead of developing a completely new and separate weapon system, the MPRS can be attached to and used with any 40-mm low-velocity (LV) grenade launcher, such as the M203.[106] The MPRS was designed by Israel Military

---

[104] "ADD/Daewoo K11 Dual-Caliber Air-Burst Weapon," discussion in Guns Corner forum, *Pakistan Defence*, April 21, 2014.

[105] "UAE Order 40 K11 Airburst Rifles from South Korea," *Asian Defence*, 2010.

[106] "IMI to Present Products and Capabilities Suitable for the Colombian Market at F-AIR 2013," *Defense-Aerospace.com*, July 2013.

Industries and attaches to existing weapon systems as an advanced sight. The device identifies the range to the target, which is said to also be able to communicate that range to other devices within the squad, and allows the soldier to set a range of fuse options not apparently available on weapons such as the K11. These different fuse settings allow for the detonation of the 40-mm grenade at a designated height above a target, point detonation, or a delay that allows the grenade to punch through softer targets, such as windows, before detonating.[107]

Figure 7.12. Israeli-Made Multi-Purpose Rifle System

SOURCE: Publicity photo from Israel Military Industries.

NOTE: The rifle system can be employed with existing 40-mm grenade launchers to provide air-burst, point detonation, or delayed fuse settings.

Air-Burst Hand Grenade

Another approach to engaging targets in defilade is the Swedish air-burst hand grenade. Considering standard hand grenades to be deficient in situations in which small obstacles often reduced a grenade's effectiveness or multiple fragments were sent into the air away from targets, the Swedish army has adopted a more efficient and effective system. Handled and thrown as an ordinary hand grenade, these air-burst grenades roll on the ground until they come to a stop and then deploy a small device that pops the grenade into the air 1.5 to 2 meters before exploding. In addition to exploding above small obstacles, the fragmentation is directed downward, toward the target, making it more lethal within its intended employment radius and less dangerous to noncombatants outside of that area. A simple manipulation of the grenade allows soldiers to instead employ it as a standard hand grenade.[108]

---

[107] Arie Egozi, "i-HLS Test: The IMI Multi Purpose Rifle System," *Israel's Homeland Security*, December 20, 2012.

[108] Sweden to Field Air Burst Grenade in 2011," *Soldier Modernisation*, Vol. 6, January 2011.

Russian and Chinese Solutions

Russia and China have traditionally employed similar approaches to counterdefilade targeting. Both use the concept of firing a grenade that has a smaller outer charge that is detonated upon hitting the ground, while the main charge is delayed until it reaches a certain height before exploding. The Russian VOG-25P is a 40-mm LV air-burst grenade developed in the 1980s that explodes when it reaches 0.5–1.5 meters off the ground. A more recent version, VOG-25PM, is similar in appearance and is fired from a standard 40-mm grenade launcher with a range of 400 meters.[109]

The Chinese antipersonnel round operates with a similar premise but is fired from the Type 69-1 rocket launcher or another RPG-7-type system (see Figure 7.13). As with the Russian round, a small charge is thought to make the round jump up after impacting on the ground before it detonates at approximately two meters in the air. Antipersonnel steel spheres reportedly give the weapon a lethal radius of 15 meters.[110]

Figure 7.13. Chinese Type 69 Antipersonnel Round

SOURCE: Photo shared by Israel Defense Forces via Flickr, CC BY-NC 2.0.

NOTE: The Chinese Type 69 antipersonnel round is fired from RPG-7 type systems increasing its effective range.

---

[109] "40 mm VOG-25P Caseless LV HE Air-Burst grenade," *Jane's Infantry Weapons*, last posted October 11, 2011.

[110] "Type 69 40 mm Airburst Anti-Personnel Round," *Jane's Infantry Weapons*, last posted August 19, 2010.

Common Themes and Notable Outliers

The previous examinations of weapon systems, squad organizations, and soldier modernization programs include both common themes and notable outliers that can be combined and potentially harvested for U.S. Army equipping ideas.

The average size of a rifle squad is just over nine soldiers, and a Marine Corps squad cannot be larger than 13. Some of the concern for keeping squads at a smaller, more manageable size has traditionally included a squad leader's ability to control the unit. A common theme of nearly every soldier modernization program is improvements in C4I systems that not only better inform soldiers and small-unit leaders but also improve their ability to communicate and monitor members of their unit. While a shared, timely, and accurate COP is continued to be sought through equipping solutions, the consequences for squad employment and training opportunities may result in decreasing the concern over squad leaders controlling more than an eight-man squad.

An important issue for mechanized infantry is IFV capacity. The average capacity of 46 IFVs from around the world is 8.28, which is smaller than the average squad size by nearly an entire person. Thus, vehicles tend to be built too small to accommodate even an average full-strength squad, let alone one that is likely to be employed with attachments. Arguably, IFVs and other transport vehicles should be designed and built around the squad as a system as opposed to designing or breaking up the squad to accommodate a vehicle. Other important vehicle characteristics can be affected by increasing troop capacity requirements, but if the squad is to be considered a system in itself, there may be detrimental effects to its employment by not accounting for how it interacts with all of its equipment, including vehicles. The Army may find value in the joint testing of systems with the USMC's Gruntworks facility, which is designed to test and optimize the human interaction with equipment and vehicles.

Two common themes in soldier and squad modernization programs are concerns for weight and power production and management. Military-developed battery technologies and advanced materials may offer future solutions, but many of the countries we reviewed seem to leverage systems common in civilian sectors, such as Bluetooth technology, PDA (personal digital assistant) displays, and digital magnetic compasses linked to GPS. These types of systems, in addition to sensors on sights and optics, are typically linked to a communication system that shares voice, data, and potentially video through its own network as well as to higher command and control systems. Automated reporting for unit position and status may simplify reporting procedures and attempts to give all leaders a shared COP.

Counterdefilade capabilities have existed for years within the Russian and Chinese militaries. More recently, other approaches to provide infantry squads with this capability have been adopted. Some, like the South Korean K11, represent similar approaches to the XM25, while others, such as the Israeli MPRS, may present a more viable and functional addition to the current M203 grenade launcher. Sweden's air-burst hand grenade also provides soldiers with an additional option of how to employ hand grenades, which have not largely changed in function or form for decades.

Advanced helmets seek to improve or facilitate all fundamental functions and senses of soldiers in combat. Soldiers with these systems can see better through day, night, and thermal fusion optics. They can hear better through integrated earphones and hearing protection. They can talk more easily through bone conduction or neck microphones. Some may be able to capture what they observe through attached light-intensifying cameras, and others may be able to control their computer systems through voice recognition software.

Attempts are being made to continue to improve soldiers' safety through reducing their visual, radar, and infrared signatures, rendering them more difficult to detect and target. Through the integration of GPS and digital compasses, warnings can be provided to prevent fratricide, and other sensors can notify the squad when they are near NBC threats, while denying the local use of RC-IEDs. Advanced navigation systems such as S-Nav, when combined with automatic GPS location reporting, will ensure that squads are confident of their geographical location and ability to navigate any terrain and route.

Sustainment improvements include training soldiers to carry and use water purification systems and foraging techniques to reduce the weight of the water that squads must carry. Similarly, integrating power sources and reducing the number and types of different batteries each soldier and squad must carry appears to be a commonly sought characteristic of most future equipping solutions. Monitoring of biological information may also assist sustainment by identifying extreme body temperatures, acceleration, and electrocardio information prior to such factors injuring a soldier.

The armies that were interviewed for this research expressed concern about the amount of weight that the typical infantryman is now carrying while dismounted from his vehicle. Considerable useful equipment has been developed over the last decade by many armies around the world—but it has to be carried. Loads that exceed 100 pounds were described as (unfortunately) typical for today's infantryman by every army that was consulted for this study.

Finally, a number of programs from around the world provide some institutional lessons that can be applied by U.S. Army equipping solutions. Ideally, the squad system would be characterized by being modular such that it can be personalized for specific billets. While some modernization programs around the world have been designed with certain versions in mind, allowing individuals to choose their own preferred display method, for example, may make training and the employment of the systems more comfortable for individual soldiers.

Modularization would also benefit Army equipping solutions as new and improved components or technologies are developed, and they could be introduced without having to update the entire system. Another important function of many squad systems is that some are designed and developed with the intent of combining with simulation systems to enhance training opportunities.

(This page is intentionally blank.)

8. Robotics

The United States has seen rapid growth in robotics technology over the last decade, and the last several years has been remarkable in the overall integration of the technology. Several thousands of systems have been added to the force structure, and the growth is expected to continue into the foreseeable future. While most of the recent integration of robotics technology has initially occurred in the airspace sector through the fielding of many new UASs, which has now gone past its tipping point, significantly more integration is likely to occur in the coming years in the ground-based sector through a wide range of UGVs. While the services have been investing in unmanned systems for several decades, the recent growth has largely been driven by congressional language that has directed the services to field unmanned systems, with specified target numbers (one-third of attack aircraft and one-third of ground combat vehicles to be unmanned).[111] While the service leadership initially balked at the target numbers at the time, by most measures, the services have met or are on the path for meeting the target numbers specified. As evidence of this trend, there are now more man-hours associated with flying UASs than combat aircraft, as measured on an annual basis, and there are more pilots trained for flying UASs than manned aircraft.[112]

While the United States currently has a lead in both air- and ground-based unmanned systems, looking at the global picture, other countries are not far behind. In some ways, unmanned systems provide a leap-ahead opportunity for countries that have not been able to field a competitive air force (top-down rationale); in other ways (bottom-up rationale), unmanned systems provide an opportunity to develop increasingly sophisticated IEDs. Whatever the rationale, there is clear evidence that the foreign market for unmanned systems will likely grow, and demand may be as high or higher than U.S. demand in the future. Some of these countries that have shown an interest or have begun development programs are not allies of the United States.

Comparison of UAS and UGV Investment and Growth in the United States

Over the last several decades, the U.S. military has been investing in the area of unmanned systems, in the form of robotics. Much of the early investment into UAVs (now referred to as UASs), have since been transitioned into the mainstream. They are so commonplace now that some argue that UASs may replace air-combat aircraft in the not-too-distant future, and some

---

[111] U.S. Congress, 106th Cong., National Defense Authorization, Fiscal Year 2001, Public Law 106-398, Washington, D.C., October 30, 2000. Appendix, Title II—Research, Development, Test, and Evaluation, Subtitle A—Authorization of Appropriations, Section 220.

[112] "Flight of the Drones: Why the Future of Air Power Belongs to Unmanned Systems," *The Economist*, October 8, 2011.

believe that the F-35 may be the last manned tactical combat aircraft that the United States Air Force, Navy, and the Marine Corps produces. In contrast to UASs, most of the UGVs have not yet transitioned into the mainstream. While much revolutionary technology exists to enable UGVs, the overwhelming majority of UGV capability still resides in the science and technology (S&T) domain.

Part of the reason for the difference can be traced back to the disparity in the investment that DoD has made in UASs and UGVs. According to recent Office of the Secretary of Defense (OSD) data from the unmanned systems roadmap, Unmanned Warfare & Intelligence, Surveillance, and Reconnaissance, DoD expenditures between FY 2007 and FY 2013 were projected to be roughly $22 billion for UASs and only $0.8 billion for UGVs.[113] Investments over the last several years for UASs and UGVs are shown in Figure 8.1. While one might argue that manned aircraft tend to be much more complex and thus more expensive platforms than ground vehicles, in the robotics realm, the opposite may actually be true. Unmanned ground vehicles that operate with a high degree of autonomy are currently seen to be much more complex than their UAS counterparts. As a result, more research, development, testing, and evaluation (RDT&E) may be needed to get to a level of proficiency.

As it stands now, according to the available data, UASs expenditures past and planned (RDT&E, production, and operations and maintenance [O&M]) have had an order of magnitude of additional funding relative to UGVs. Perhaps more pragmatically, the underlying reason for the disparity in investment is not well correlated to the complexity of the problem that each unmanned system must address. To a large extent, developing advanced UGVs for warfare has been considerably more challenging than developing UASs, since UGVs must operate in much less structured environments. However, as the technology continues to evolve, the likelihood of advanced UGVs becoming a part of the fielded forces becomes much higher. Some of the systems currently in the S&T domain will enter the development and production pipeline.

---

[113] Dyke D. Weatherington, Director, Unmanned Warfare & Intelligence, Surveillance, and Reconnaissance, personal communication, telephone, 2013.

Figure 8.1. Comparison of UAV/UAS and UGV RDT&E, Production, and O&M in Recent Years

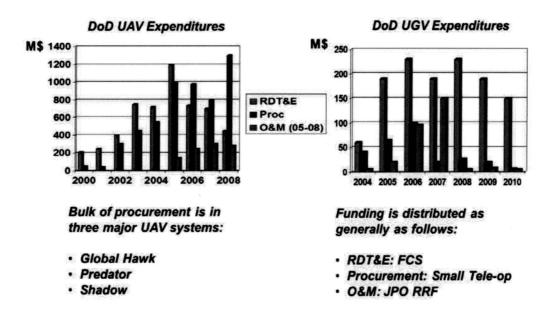

Bulk of procurement is in three major UAV systems:

- **Global Hawk**
- **Predator**
- **Shadow**

Funding is distributed as generally as follows:

- **RDT&E: FCS**
- **Procurement: Small Tele-op**
- **O&M: JPO RRF**

SOURCE: Created using data from Weatherington, 2013.

Note: FCS = Future Combat System.

Examples of U.S. and Foreign UAS Programs

Because of the relative newness of robotics technology, perhaps combined with the availability of published basic academic research, the lead that the United States has established in unmanned systems can be challenged. Essentially, there is a relatively low cost of entry into the UAS market, and the return on investment appears to be high, with short payback periods. In tactical-level UASs, both the United States and Israel are the primary manufacturers of the current platforms that are fielded. The United States has approximately 7,000 UAS platforms currently in the inventory, as compared with only 50 platforms ten years ago.[114]

In large part because of the successful UAS experience the United States has had, many other countries have since established UAS programs, including many U.S. allies as well as other countries, such as China, Iran, and Pakistan. Currently, there are 960 UASs being manufactured by 270 companies in 57 countries. As for growth, these numbers have significantly increased over the last two years; the number of UAS platforms is up 40 percent, the number of UAS companies entering the field is up 20 percent, and the number of countries has increased 50 percent.

Two key U.S. systems and representative foreign counterparts are shown in Figure 8.2. While the Army's UAV referred to as the Shadow 200, is technically a U.S. system, it originated out of a

---

[114] "Drones—Who Makes Them and Who Has Them?" *Radio Free Europe Radio Liberty*, January 31, 2012.

collaborative effort involving AAI (a subsidiary of Textron) and Israel Aircraft Industries (IAI) called the Pioneer, which was based on an Israeli design. In general, the breadth of the UAS market is quite large and growing, where most modern industrial countries either have indigenous UAS programs (such as China) or have acquired platforms from countries that build them. Additionally, some countries (Japan, for example) have a competitive advantage since they have fewer restrictions associated with UAS operation and manufacturing—for example, access and operation to the national airspace is less restricted, resulting in faster development and training times. Additionally, some relatively crude systems have been crafted out of spare airplane parts in third world countries.

Figure 8.2. Notable UASs

Pioneer (Israeli and U.S.)

Raven (U.S.)

Shahed 129 (Iranian)

Sky Warrior (U.S.)

SOURCES: (clockwise from top left) photo by the U.S. Navy, photo by the U.S. Army, photo courtesy of Sepah News, and photo by Sergeant Travis Zielinski, 1st ACB, 1st Cav. Div., USD-C.

Of particular note is Iran's interest in UASs. Figure 8.2 shows the Islamic Revolutionary Guard Corps (IRGC) Shahed, or Witness, 129. Unveiled in 2012, this UAS is reportedly capable of traveling more than 2,000 km with more than 24-hour endurance. This UAS can be armed and can be used for surveillance and combat missions; it was developed indigenously.

It is clear that the Iranian military has interest in further developing and promoting its UAS capability. Prior to the unveiling of the Shahed 129, the Iranians had developed a tactical UAV/UAS called the Ababil, or Swallow, which was operational since the mid-1980s in various forms. From this initial experience in technology, the field has grown considerably, both in number of platforms and in overall capability. The bomber drone referred to as the Karrar, or Striker, was unveiled in 2010. And in March 2013, Iran unveiled a tactical UAS referred to as the Hamaseh, or Valiance.

Recently, the Iranian leadership also announced the existence of four other UAS programs, Azem-2, Mohajer B, Hazem 3, and Sarir H110. While perhaps not competitive with U.S. (and Israeli) UAS capability at this moment, the surge in the number of programs and the overall growth trajectory should make this capability of key interest to decision- and policymakers.

China is also rapidly developing and continuing to grow an indigenous UAS capability. Two examples of this effort are shown in Figure 8.3. The Pterodactyl has a similar shape and similar specifications to the Predator the Chinese have been developing a stealth (based on appearance) UAS referred to as the Sharp Sword. In some ways, this shape resembles different concepts originating out of the U.S. unmanned combat air vehicle program. Since the Sharp Sword's release in November 2013, much still is unknown (as of this writing).

Figure 8.3. Examples of Recent Chinese UAS Development

Pterodactyl (China)

Sharp Sword (China)

SOURCES: Photos by Baiweiflight, CC BY-SA 3.0.

While the U.S. Army is likely reliant on the Air Force or Navy for evolving the UAS capability, it will likely user the future capability well into the future.

Status of UGV Development and Production

In comparison with UAS investment, the United States has been much slower to invest in, develop, and ultimately produce UGVs for fielding. As noted earlier, recent estimates suggest that UGV investment is about an order of magnitude behind that of the UAS investment. Although many have argued that UGV technology is mature, it continues to reside largely within the S&T base—perhaps in part because of the lack of adequate policy and doctrine. Additionally, depending on the particular use of the platform, UGVs can be a much more sophisticated, complex operation than UAS operations. While there has been enormous success in the use of UGVs for combating IEDs in Iraq and Afghanistan, these systems are composed of a relatively low level of technology, where they are controlled through teleoperation. The manpower required to operate these counter-IED robots tends to be as high as if the operation were performed as a manned operation (the primary benefit is the reduction of casualty risk).

Many next-generation UGV systems with higher levels of autonomy already exist and are being tested and used in various experimental capacities. A few of the more notable UGV systems are shown in Figure 8.4. The Mobile Detection Assessment and Response System (MDARS), which was developed by the Army Research Laboratory, has existed for more than a decade in various forms. It is approximately a one-ton platform, and it was intended for unmanned base or perimeter security. MDARS has been used in limited roles since an initial fielding of several platforms, and a weaponized version of this platform was built by Space and Naval Warfare Systems Command (SPAWAR). In contrast, the unmanned ground combat vehicle PerceptOR Integrated Crusher was a Defense Advanced Research Projects Agency (DARPA)–funded initiative and was built by Carnegie Mellon University's National Robotic Engineering Center (NREC). This platform was envisioned to be a much larger and heavier UGV, at six tons, and with higher levels of mobility and combat capability.

Figure 8.4. Notable Tactical UGV Platforms That Have Undergone Extensive Field Testing and Evaluation or Fielding

Guardium (Israeli)

MDARS (U.S.)

Crusher (U.S.)

SOURCES: (Clockwise from top left) publicity photo from G-NIUS Unmanned Ground Systems, publicity photo from General Dynamics, and publicity photo from Carnegie Mellon's National Robotics Engineering Center.

As with UASs, the Israeli military has significant UGV capability; perhaps even more notable, this capability has already been fielded in the Israeli Army. The Guardium UGV, developed by the Israeli company G-Nius, is a joint venture between Israel Aerospace Industries and Elbit Systems. It has been adapted for a range of terrain types, and it can operate semiautonomously for several days. It can travel 50 km per hour and has a payload of approximately 300 kg. Payloads can consist of an Explosive Ordinance/infrared camera, Hostile Fire Indicator (HFI), a Missile Approach Warning System (MAWS), a laser warning system, a two-way audio link, chemical snifters, and a radio frequency identification (RFID) interrogator. It is also possible to install a remotely operated weapons system as well as nonlethal weapons suites. Other robot-related systems have been fielded in the recent past. One notable example of this is the South Korean SGR-1, which has been deployed to the demilitarized zone (DMZ); see Figure 8.5. This sentry system is built by Samsung Techwin and is generally designed to provide fire support in order to include 5.5-mm machine guns and 40-mm grenade launchers in the DMZ.[115] While this system's fire control is ultimately overseen by a human, the surveillance and reconnaissance capability is fully automated. Thus, it autonomously searches and identifies candidate targets for human controllers, who ultimately decide whether or not to engage the candidate targets.

Figure 8.5. A South Korean Automated Sentry, the Samsung SGR-1, Deployed in the DMZ

SOURCE: Publicity photo from Murich Teknoloji.

Many more UGVs are in the pipeline, including a wide range of logistics and support systems. This is a UGV area that has been well developed for many years and yet has not been fielded for quite some time. The Army's Tank Automotive Research, Development, and Engineering Center (TARDEC) has built, tested, and demonstrated semiautonomous UGVs for use in a range of Army missions, many of them support functions, such as convoy capability. A decade ago, Stryker vehicles were converted to semiautonomous platforms and operated in autonomous, leader-follower mode for several hours without manned intervention. Since that demonstration, there have been many other platforms that have been developed with the goal to support ground forces. A few of these key programs are shown in Figure 8.6, which include dismounted infantry support systems, such as the Squad Mission Support System (SMSS), and the Big Dog and combat logistics patrol (CLP) capabilities, such as the autonomous mobility appliqué system (AMAS).

---

[115] Timothy Hornyak, "Korean Machine-Gunrobots Start DMZ Duty," *CNET*, July 14, 2010.

Figure 8.6. Examples of the Support and Logistics UGVs Being Developed

Squad Mission Support System

Big Dog

Autonomous Mobility Appliqué System

SOURCES: (Clockwise from top left) publicity photo from Lockheed Martin, publicity photo from DARPA, and publicity photo from Lockheed Martin.

The SMSS, built by Lockheed Martin, is a helicopter-transportable robotic platform that can self-navigate through soldier following, waypoint specification, and GPS navigation. As for payload and range, it can carry half a ton 125 miles downrange.[116] The Big Dog (the latest version is called Alpha Dog) is built by Boston Dynamics. The Big Dog is intended to provide support to dismounted soldiers over very difficult terrain. The Big Dog can carry 340 pounds 12 miles downrange (Alpha Dog can carry 400 pounds for 20 miles), and it can travel up to speeds as high as 25 miles an hour.

The AMAS, also built by Lockheed Martin, is intended to be an optionally manned vehicle add-on capability. Essentially, by taking drive-by-wire vehicles and adapting different kits to include sensor packages and processing alternatives, existing logistics platforms can be converted to operate semiautonomously. While the current focus is to augment drivers and assistant drivers

---

[116] John Reed, "Army Fielding Robo Jeeps in A'Stan," *DefenseTech.org*, August 5, 2011.

operating as part of a CLP, largely to reduce accidents and free up attention to search for IEDs, the long-term prospect for this technology can be crew or manpower reduction.[117]

Other Types of Robotics Capability to Consider

In addition to the traditional UGV programs, there are other robotics and robotics-related initiatives under way that can revolutionize the way that things are conducted in the military. These include bipedal robots, exoskeletons, and commercial UGVs; key examples of these classes of robotics are shown in Figure 8.7.

Bipedal Robots

Although the United States has largely focused on wheeled UGVs, there has been tremendous interest in bipedal robots, mostly coming from Japan. For well over a decade, Japanese companies have been developing "humanoid" robots, largely for health and geriatric care for their aging population. While these robots are not necessarily intended for military functions, it is possible that downstream they could be adapted. One of the key benefits here is that these robotic systems have inherent flexibility in movement and mobility, such that they parallel human dexterity. Wheeled UGVs do not have this same degree of flexibility.

Exoskeletons

Both Japan and the United States have been evolving the capability of exoskeletons. Again, the Japanese have been developing the capability largely to assist with elderly mobility; whereas much of the U.S. focus is military based, specifically on enhancing individual soldier performance.[118] This technology has been relatively slow to develop, given the complexity of interpreting user-specified signals and developing appropriate actions and the limitations of the onboard power supply to meet a 24–72-hour operational mission profile.

Commercial UGVs

Recently, Google announced that it was entering into the robotics market by commercializing an autonomous car through the Google Self-Driving Car initiative. Google's motivation is to develop the capability as a substitute for human driving, thus freeing up humans to perform other functions while being transported. The potential market size is enormous, about 30 billion man-hours per year. While Google's entry into the robotics market was dramatic and the progress they've since seen has been even more impressive, it built on much of the technology that the military, including the DARPA grand challenges, has been pursuing for some time.

---

[117] John Matsumura et al., *Assessing the Impact of Autonomous Robotic Systems on Army Force Structure*, Santa Monica, Calif.: RAND Corporation, RR-226-A, 2012.

[118] Much of the U.S. capability is being developed by SARCOS, based in Utah.

Figure 8.7. Other Robotics Initiatives That Can Impact the Military: Bipedal Robotics, Exoskeletons, and Commercial Initiatives

Bipedal Robot (Japanese)

Exoskeleton (U.S.)

Self-Driving Car (U.S.)

SOURCES: (Clockwise from top left) publicity photo from Honda, publicity photo from Lockheed Martin, and photo by Steve Jurvetson, CC BY 2.0.

Conclusions

Although the United States has a robust, well-developed, and wide-spectrum UAS and UGV capability, it is clear that there are foreign competitors that have specialty areas or niches within the broad field. Since much of the space within future unmanned system capability has yet to be defined, it is not clear how vulnerable the U.S. lead in robotics is. For UASs, the United States maintains a lead across the broad front, but other countries, including Israel and China, are not far behind, and in some particular areas might be ahead. Iran appears to be putting a large number of eggs in the UAS basket as well.

For UGVs, the technology needed for successful use is much more complex. Experts have argued that UGVs are about 10 to 15 year behind in development and production. If this is the case, there will be a burgeoning market five to ten years from now. For UGVs, the U.S. focus has largely been on wheeled platforms, with the most recent push being for optionally manned vehicles. This strategy is very conservative, and can be surpassed by leap-ahead initiatives that focus on smaller autonomous systems that do not have to have man-in-the-vehicle (or man-in-the-loop, for that matter) for function. The Israeli fielding of Guardium is an example where tactical and operation experience has been gained that extends beyond the U.S. tactical experience, largely limited to teleoperated counter-IED robots. Also, if interchangeability with high levels of dexterity between soldiers and robots (bipedal) is a long-term goal, the United States does not currently have the lead.

Perhaps most important to note is the fragility of the U.S. position in robotics. Generally, basic robotics technology is available, and if modest resources are devoted, it is possible to build competency relatively quickly. The Google Cars experience is a case in point. Essentially, there are relatively low barriers for entry, and the learning curve for competency is still comparatively flat, relative to other advanced technologies. As robotics capability progresses and operations become even more complex, this may change. While there is a tendency to assume that the technology will progress at a pace that the United States dictates, this will only be the case only if the U.S. policymakers and military leaders actively manage this technology area.

9. Conclusions and Recommendations

This research compared current and near-future U.S. Army systems and capabilities with those of selected other armies and the U.S. Marine Corps. Some of the militaries that were used for the comparisons are allies of the United States, while others are potential competitors. In terms of overall capability, the U.S. Army remains in a league of its own—no other army in the world has the same depth and breadth of capabilities. That said, there are areas where capability gaps are appearing due to the modernization plans of other armies. These are areas that the U.S. Army should consider addressing. Additionally, there are without question a number of ideas from other armies that are worthy of consideration.

A lot of ground was covered in this research, yet RAND did not have the ability to conduct a compressive comparison of U.S. Army systems and their foreign counterparts in all the warfighting functions. This final chapter will highlight areas that the Army should give special consideration to. It is important to note that the insights from this research stress similar-system-to-similar-system comparisons. It was beyond the scope of this research to compare and contrast, for example, long-range rocket launchers and aircraft.

Indirect Fires

The Trends in Long-Range Multiple Rocket Launchers Are a Cause for Concern

The indirect fires chapter showed the trends in the range and accuracy in foreign MRLs. Easy to produce, tactically mobile, and with ranges that are now typically well over 100 km, the long-range heavy MRL has a major Anti-Access/Area-Denial capability if present in an opponent's arsenal. Today's heavy MRLs are now capable of attacking targets at ranges that in the 1990s were the purview of short-range ballistic missiles. The accuracy of these weapons varies considerably (fairly high with the Chinese WS-2, low in the case of the Iranian Fajr-5), but even MRLs with relatively poor accuracy can threaten area targets, such as airfields, where fragile, and very expensive, aircraft could be damaged or destroyed on the ground by barrages of rockets. If the United States does not have the capability to counter this growing antiaccess threat, future operations will be much more challenging. It should be noted that the long-range MRL threat is also figuring prominently in other RAND work for the Army, such as ongoing antiaccess studies for TRADOC and Headquarters, Department of the Army.

Given that the trend in the foreign heavy MRLs is toward ranges of well over 100 km, the current U.S. Army MLRS family of munitions (including GMLRS, with a range of 84 km) is falling considerably behind. Additionally, the Army's latest Firefinder radar capability could also lack the range required to find hostile long-range heavy MRLs. Some combination of Army and joint detection-defensive-offensive capabilities is needed to mitigate this threat. Actions that the Army should consider include:

- Ensuring that an ATACMS-like capability is retained, permitting a quick-response, surface-to-surface counterfire capability at ranges of 300 km (the range of today's extended-range ATACMS) or beyond. It is unlikely that even orbiting aircraft could be as responsive as an organic Army missile capability in terms of their ability to engage a just-identified enemy MRL. The Fires Center of Excellence plans for GMLRS Increment IV may or may not be adequate as a replacement for ATACMS, depending on the range and payload that the new rocket is able to achieve.
- Explore Army and joint weapons detection capabilities to quickly locate the firing positions of long-range heavy MRLs that could be firing from ranges of 200 km or greater. Due to the extended ranges of modern heavy MRLs, some type of airborne weapons locating sensor will probably be needed.

Increase the Munitions Available for the MLRS and HIMARS

When compared with their foreign counterparts, the Army's rocket launchers have a limited suite of warhead options. Currently, DPICM and HE warheads are the only munitions that MLRS and HIMARS can fire. The 2019 submunitions limitations mean that the Army will have to replace most of its DPICM warheads. Given that reality, there may be an opportunity to examine other warhead options of the type that are used in other armies' MRLs (guided submunitions, fuel-air explosive, etc.).

Movement and Maneuver

Trends in World IFVs are Toward Heavier-Caliber Cannons and Remote Weapons Stations

The 25-mm Bushmaster on the Bradley IFV is a successful, powerful weapon. When compared with world trends, however, the Bradley's gun is falling behind. The clear trend is toward 30–40-mm weapons on IFVs and armored personnel carriers, whether tracked or wheeled. In that regard, the .50-caliber machine gun on the infantry carrier version of the Stryker is far behind its foreign counterparts.

The Army is currently developing the Ground Combat Vehicle as a replacement for the Bradley infantry carrier. If the development of the GCV is successful, mounting an automatic cannon of at least 30 mm would be appropriate. If the GCV does not come to pass, or is significantly delayed, the Army should consider rearming the Bradley with a heavier automatic cannon. A heavier cannon of 30–40 mm would increase the ability of the Army's infantry fighting vehicles to engage other armored vehicles as well as buildings. Given the ever-increasing global trend toward urbanization, the latter capability would be significant.

In addition to considering a heavier gun for Bradley and/or GCV, the use of the RWS should also be examined. A number of the new foreign IFVs and APCs (tracked and wheeled) are using the RWS and remote turrets. Given improvements in vehicle video systems, today there is a much better ability for a vehicle crew to achieve high levels of situational awareness without necessarily needing a manned turret. The RWS has the additional advantage of lowering vehicle weights and possibly increasing the internal volume available inside the AFV.

Other Militaries Are Employing Combinations of Unmanned Aerial Systems and Attack Helicopters in Reconnaissance and Scouting Roles.

The U.S. Army added weapons to the OH-58 Kiowa years ago, thus increasing the overall utility of that aircraft. In Iraq and Afghanistan, OH-58s were used for scouting as well as roles such as convoy escort, where their light armament was useful. That said, in other armies and air forces (in some other militaries, helicopters are controlled by their air forces), there is little evidence of purpose-built observation and reconnaissance manned helicopters. Combinations of attack helicopters and UASs, as well as light utility helicopters, are being used in those roles.

While other militaries lack the capability that the Army's OH-58s provide, they are nevertheless saving money by avoiding a purpose-built aircraft for that role. The ever-increasing capability of UASs is contributing to the lack of interest in specialized, manned, and reconnaissance helicopters.

The U.S. Army should carefully examine the need for a new specialized manned reconnaissance aircraft in light of global trends and the increasing ability of combinations of attack helicopter and UASs to perform the scouting, observation, and reconnaissance functions.

Protection

Examine What the Best Defensive Systems Are for Countering Incoming Long-Range Rockets

The recommendation to examine the best defensive systems for countering long-range rockets is due to the trend in heavy MRLs that was highlighted in the fires section. Given that shooting multimillion-dollar Patriot missiles at a barrage of incoming rockets is not feasible except in extreme circumstances, various gun, missile, and directed energy defensive systems should be examined.

Ideally, these defensive systems should be easily deployable so that they can be rapidly moved to a just-seized port or airfield to establish a defensive capability. Indeed, from the perspective of XVIII Airborne Corps, the Army's primary contribution to the global response force, there would be a critical requirement to have rapidly deployable defensive systems (able to deploy in as few U.S. Air Force transport aircraft as possible) that can be quickly set up to defend a just-captured airhead from incoming rocket fire.

In addition to the active defense options mentioned, electronic countermeasures and decoys should also be explored. The guidance systems of long-range rockets may be vulnerable to electronic countermeasures, especially if they have onboard guidance systems for in-flight course correction.

Squad Comparison

Armies Are Aware That Today's Dismounted Infantryman Has Become Grossly Overloaded

The French, British, Australian, German, Canadian, and Israeli army representatives that RAND interacted with all said that dismounted infantrymen are overloaded. The foreign army representatives mentioned that while the new equipment that has become available to the infantry in the last decade is impressive (better body armor, new weapons, counter-IED electronics, computers, etc.), the cumulative effect of all the new *kit* (to use the phrase of the British Army representatives) has been to encumber the typical infantryman with 60–100 pounds (or more) of weapons, ammunition, and other equipment. Without exception, the representatives of the other armies consulted in this study said that not only does this trend have to be stopped, it needs to be reversed.

The U.S. Army should consider approaches to deburden infantrymen. There are various ways to do this, including being rather ruthless in establishing and enforcing rules as to what the dismounted infantryman should normally be expected to carry. There are also technological solutions for light infantry. These include adding judicious numbers of small vehicles that can be used to carry part or most of the load of a squad. In addition to small manned vehicles, increasing the number of robotic carrying systems would also allow much of the load to be removed from the typical infantryman. Humanoid-like robots would be more appropriate in buildings and in other restrictive terrain than small vehicles.

Squad Size Varies Considerably, Depending on the Army, as Does the Troop Carrying Capacity of IFVs and APCs

The squads of other armies that were examined in this research varied from 8 to 12 personnel. If the USMC rifle squad is included, the largest squad was 13 personnel. Some armies include light machine guns and antiarmor weapons in their rifle squads, while others place those weapons at the platoon level. In some cases, foreign armies divide their squads into teams of three to four personnel who have either identical or different tactical functions. That said, the basic mission and capabilities of foreign rifle squads are similar to those of the U.S. Army.

One of the reasons for different squad sizes is the constraints imposed by the IFVs or APCs that are used to transport and support the squads. In the case of the German Army, two different squad sizes are used: nine personnel for light infantry, and seven in the case of *panzergrenadiers*, which are organic to armored units. The smaller squad in the case of the *panzergrenadiers* is due to the carrying capacity of the older Marder or new Puma IFV.

No foreign army that was examined in this study used more than one vehicle to transport an infantry squad, although some were willing to reduce their squad sizes in order to fit the existing or planned IFVs or APCs. As the U.S. Army considers its future options for the new GCV and the older Bradley, and their associated infantry squads, there is no clear trend in how other armies man and equip their squads.

Robotics

While the U.S. Army Retains an Overall Lead in Military Robotics, a Number of Other Militaries Are Currently Ahead in Selected Areas in the Robotics Field, and Technology Breakouts by a Number of Countries Are Possible

The Army's use of unmanned ground systems increased considerably during the fighting in Iraq and Afghanistan; unmanned systems were very useful to find and disarm IEDs, for example. Despite the experience gained in recent conflicts, there are a number of important areas within the field of military robotics where the United States does not lead.

The use of robotic vehicles for both reconnaissance and surveillance is increasing. Rapid advances in technology are enabling increasing levels of autonomy for these systems, as well as improving the payloads and speed of unmanned ground systems. In some militaries, notably the Israeli Defense Force, the use of armed robotic vehicles is increasing. That is an area where foreign militaries are currently ahead of the U.S. Army. To some extent this is due to policy considerations—thus far U.S. decisionmakers have been hesitant to permit armed robotic systems, especially for missions where relatively high levels of autonomy would be needed.

In the area of bipedal, walking, humanoid-like robots, others are clearly ahead of current U.S. capabilities. Japan, in particular, leads this field. As mentioned in the squad comparison section, the use of walking, bipedal robots could significantly increase the capability of dismounted infantry, where much of the load that the soldiers carry could be moved to robots who would be able to move with the squad among buildings or among restrictive terrain, where even small vehicles could not go.

The robotics field is currently wide open; new military entrants could quickly achieve a relatively high level of capability by capitalizing on civilian robotics research. This raises the possibility of breakouts by possible competitors, who could, after a few years of intense effort, pull significantly ahead of the U.S. military in selected areas.

The bottom line is that in the rest of the world, there is a clear move toward military robots. Therefore, the Army should, to the extent possible, support research and development in this important new technology field with the goal of improving its capabilities, as well as gaining and maintaining technology leadership.

(This page is intentionally blank.)

Bibliography

"40 mm VOG-25P Caseless LV HE Air-Burst Grenade," *Jane's Infantry Weapons*, last posted October 11, 2011. As of June 23, 2014:
https://janes.ihs.com/CustomPages/Janes/DisplayPage.aspx?DocType=Reference&ItemId=+
++1360280&Pubabbrev=JIW_

"155 mm howitzer M198," *Jane's Armour and Artillery*, updated March 12, 2012.

"155mm V-LAP Round," *Jane's Ammunition Handbook*, updated January 25, 2013.

"333 mm Fadjr-5 Iranian Rocket," *Jane's Armour and Artillery*, updated February 5, 2013.

2014 Army Equipment Modernization Plan, Headquarters, Department of the Army, May 13, 2013. As of June 23, 2014:
http://www.g8.army.mil/pdf/AEMP2014_lq.pdf

"AAV-7," *Marines.com*, undated. As of June 23, 2014:
http://www.marines.com/operating-forces/equipment/vehicles/aav-7#features

"ADD/Daewoo K11 Dual-Caliber Air-Burst Weapon," discussion in Guns Corner forum, *Pakistan Defence*, April 21, 2014. As of June 23, 2014:
http://defence.pk/threads/add-daewoo-k11-dual-caliber-air-burst-weapon-south-
korea.310317/

Advanced Artillery System: SIAC 155/52, brochure, Madrid: General Dynamics European Land Systems, January 2012. As of August 14, 2014:
http://www.gdels.com/brochures/artillery_artillery.pdf

"AH-1W/AH-1Z Super Cobra Attack Helicopter, United States of America," *Army-Technology.com*, undated. As of June 23, 2014:
http://www.army-technology.com/projects/supcobra/

"Airbus Helicopters EC 225 and EC 725," *Jane's Defence Equipment and Technology*, February 1, 2013. As of July 18, 2014:
https://janes.ihs.com/CustomPages/Janes/DisplayPage.aspx?DocType=Reference&ItemId=+
++1342585&Pubabbrev=JAWA

Air Force Tactics, Techniques, and Procedures (AFTTP) 3-1, "Rotary-Wing Aircraft, Employment, and Tactics," in *Threat Guide: Threat Reference Guide and Countertactics*, pp. 8–56, December 3, 2012.

"Amphibious and Special Forces—China," *Jane's Amphibious and Special Forces*, November 13, 2002.

Amos, James, *Commandant of the Marine Corps: 2013 Report to the House Armed Services Committee on the Posture of the United States Marine Corps,* House Armed Services Committee, April 16, 2013. As of June 23, 2014:
http://docs.house.gov/meetings/AS/AS00/20130416/100659/HHRG-113-AS00-Wstate-AmosUSMCG-20130416.pdf

Andrew, Martin, "PLA Mechanised Infantry Division Air Defence Systems: PLA Point Defense Systems," technical report APA-TR-2009-0301, *Air Power Australia*, March 2009, last updated January 2012. As of June 24, 2014:
http://www.ausairpower.net/APA-PLA-Div-ADS.html

"Anglo-American Effort Improves British Logistic Vehicles," *Defense Update Magazine*, June 2009. As of June 24, 2014:
http://defense-update.com/features/2009/april/british_vehicle_protection_060409.html

"AN/PSS-14 Handheld Standoff Mine Detection System (HSTAMIDS)," *GlobalSecurity.org*, undated. As of June 23, 2014:
http://www.globalsecurity.org/military/systems/ground/hstamids.htm

Antipersonnel Mine M18A1 and M18, Field Manual 23-23, Headquarters of the Department of the Army, January 1966. As of June 23, 2014:
http://www.globalsecurity.org/military/library/policy/army/fm/23-23/

"Australian Light Armoured Vehicle," Australian Army, undated. As of June 23, 2014:
http://www.army.gov.au/Our-work/Equipment-and-clothing/Vehicles/ASLAV

Bacon, Lance M., "Stryker Gets Another Round of Upgrades," *The Army Times*, last updated August 6, 2012. As of August 6, 2013:
http://www.armytimes.com/article/20120806/NEWS/208060321/Stryker-gets-another-round-upgrades

"BAE Systems, Global Combat Systems 155mm Lightweight Howitzer (M777)," *Jane's Armour and Artillery*, updated March 12, 2012.

"BAE Systems M2 Infantry Fighting Vehicle/M3 Cavalry Fighting Vehicle," Jane's Armour and Artillery, last updated, January 14, 2013. As of July 9, 2014:
https://janes.ihs.com/CustomPages/Janes/DisplayPage.aspx?DocType=Reference&ItemId=+++1501113&Pubabbrev=JAFV

"BAE Systems US Combat Systems M109A6 155 mm Paladin Self-Propelled Howitzer," *Jane's Armour and Artillery*, updated February 7, 2012.

Bell Helicopter, *H-1 Program: AH-1Z and UH-1Y*, No. 1, 2012–2013.

"Bell Helicopter AH-1Z Earns Navy Approval for Full Rate Production," *Shephard News*, December 10, 2010. As of June 23, 2014:
http://www.shephardmedia.com/news/rotorhub/bell-helicopter-ah-1z-earns-navy-approval-for-full-rate-production/7907/

"The Bell AH-1Z Zulu," BellHelicopter.com, undated. As of June 23, 2014:
http://www.bellhelicopter.com/Military/AH-1Z/1291148375494.html#/?tab=features-tab

Bledsoe, Sofia, "Team Chinook Signs CH-47F MYII Contract; Cost Savings of $810 Million," *Army.mil*, June 14, 2013. As of June 23, 2014:
http://www.army.mil/article/105654/Team_Chinook_signs_CH_47F_MYII_contract-cost_savings_of810_million/

"BM-30 Smerch," *Military-Today*, undated. As of June 23, 2014:
http://www.military-today.com/artillery/smerch.htm

"BMD-4M Airborne Armoured Infantry Fighting Vehicle," *Army Recognition*, undated. As of June 23, 2014:
http://www.armyrecognition.com/russia_russian_army_light_armoured_vehicle_uk/bmd-4m_airborne_armoured_infantry_fighting_vehicle_technical_data_sheet_specifications_pictures.html

"Boeing Awarded US Army Contract for 14 Additional CH-47 Chinook Helicopters," news release, *Boeing.com*, January 11, 2012. As of June 23, 2014:
http://boeing.mediaroom.com/index.php?s=43&item=2099

"Boeing, US Army Mark Delivery of 1st AH-64D Apache Block III Combat Helicopter," news release, *Boeing.com*, November 2, 2011. As of June 24, 2014:
http://boeing.mediaroom.com/index.php?s=43&item=2000

British Land Forces, *British Army: Vehicles and Equipment*, United Kingdom Ministry of Defence, 2012.

Butler, Amy, "U.S. Army Prepares for Full-Rate AH-64E Production," *AviationWeek.com*, October 26, 2012. As of June 23, 2014:
http://www.aviationweek.com/Article.aspx?id=/article- xml/asd_10_26_2012_p03-01-511015.xml

"Canadian Army Support Vehicles," presentation by Major General Ian Poulter Canadian Army, February 6, 2012.

"CH-47F Chinook Backgrounder," *Boeing.com*, March 2012. As of July 9, 2014:
http://www.boeing.com/paris2013/pdf/BDS/Bkgd_CH-47F_0613.pdf

"CH-47F Chinook Helicopter," U.S. Army, September 21, 2011. As of June 23, 2014:
http://www.army.mil/media/220640

"CH-47F Improved Cargo Helicopter (ICH)," *Globalsecurity.org*, July 2011. As of June 23, 2014:
http://www.globalsecurity.org/military/systems/aircraft/ch-47f-ich.htm

"CH-53K: The U.S. Marines' HLR Helicopter Program," *Defence Industry Daily*, May 6, 2014. As of June 23, 2014:
http://www.defenseindustrydaily.com/ch53k-the-us-marines-hlr-helicopter-program-updated-01724/

"CH-53X Super Stallion," *Globalsecurity.org*, undated. As June 23, 2014:
http://www.globalsecurity.org/military/systems/aircraft/ch-53x.htm

"Chapter Five: Russia and Eurasia," *The Military Balance*, Vol. 113, No. 1, 2013, pp. 199–244. As of June 24, 2014:
http://www.tandfonline.com/toc/tmib20/113/1

"Chapter Four: Europe," *The Military Balance*, Vol. 113, No. 1, 2013, pp. 89–198. As of June 24, 2014:
http://www.tandfonline.com/toc/tmib20/113/1

"Characteristics," Tiger, *Eurocopter.com*, undated. As of June 24, 2014:
http://www.airbushelicopters.com/site/en/ref/Characteristics_190.html

"Chinese Z-10," *Global Military Review*, undated. As of June 23, 2014:
http://globalmilitaryreview.blogspot.com/2011/04/chinese-z-10-gunship-helicopter.html

Clements, Paul, and John Bergey, *The U.S. Army's Common Avionics Architecture System (CAAS) Product Line: A Case Study*, Carnegie Mellon Software Engineering Institute Technical Report, September 2005.

"Combat Vehicle 90 (CV90) (Stridsfordon 90) Infantry Fighting Vehicle," *Jane's Land Warfare Platforms: Armoured Fighting Vehicles*, updated November 21, 2011.

"COMFUT," Deagel.com, undated. As of June 23, 2014:
http://www.deagel.com/Land-Warriors/COMFUT_a000014001.aspx

"ComFut: Options to Address," *Soldier Modernisation*, Vol. 6, January 2011. As of June 23, 2014:
http://www.soldiermod.com/volume-6/comfut.html

"Common Avionics Architecture System (CAAS)," *Rockwellcollins.com*, undated. As of July 19, 2014:
http://www.rockwellcollins.com/Capabilities_and_Markets/Air/Rotary_Wing/Common_Avionics_Architecture_System.aspx

"Common Remotely Operated Weapon Station (CROWS)," in Federation of American Scientists, *United States Army Weapons Systems 2013*, United States Army, pp. 78–79. As of August 5, 2013:
http://fas.org/man/dod-101/sys/land/wsh2013/78.pdf

"CPJ COMPACT-R Convoy Protection Jammer with Smart Responsive Jamming Technology," *Jane's Defense & Security Analysis*, March 4, 2013.

"The Crusher—DARPAs Six Wheeled Autonomous Robot," *HighTech-Edge.com*, March 14, 2008. As of June 23, 2014:
http://www.hightech-edge.com/crusher-darpa-autonomous-robot-iphone-xbox-controller/1417/

Denel, "155mm V-LAP Round," *Jane's Ammunition Handbook*, updated January 25, 2013.

Dent, Steve, "Rheinmetall 50kW Laser Weapon Aces Latest Test, Pew-Pews a 3-Inch Ballistic Target," *Engadget*, December 21, 2012. As of June 23, 2014:
http://www.engadget.com/2012/12/21/rheinmetall-50kw-laser-weapon-aces-latest-test/

DOMINATOR®, Enhanced Combat Effectiveness for Infantry Units and Special Forces, St. Netanya, Israel: Elbit Systems, 2009. As of June 23, 2014:
http://elbitsystems.com/Elbitmain/files/Dominator.pdf

"Draco," *Military-Today.com*. As of June 23, 2014:
http://www.military-today.com/artillery/draco.htm

"Drones—Who Makes Them and Who Has Them?" *Radio Free Europe Radio Liberty*, January 31, 2012. As of June 23, 2014:
http://www.rferl.org/content/drones_who_makes_them_and_who_has_them/24469168.html

Egozi, Arie, "i-HLS Test: The IMI Multi Purpose Rifle System," *Israel's Homeland Security*, December 20, 2012. As of June 23, 2014:
http://i-hls.com/2012/12/i-hls-test-the-imi-multi-purpose-rifle-system/

"Elbit Expands Range of Autonomous Ground Vehicles," *Defense-Update.com*, 2007. As of June 23, 2014:
http://defense-update.com/features/du-1-07/elbit_UGV.htm

"ER/MP Gray Eagle: Enhanced MC-1C Predators for the Army," *Defense Industry Daily*, April 30, 2014. As of June 23, 2014:
https://www.defenseindustrydaily.com/warrior-ermp-an-enhanced-predator-for-the-army-03056/

Eshel, Tamir, "Israeli Multi-Purpose Tank Ammo Redesigned to Fit the 125mm Gun of the T-90S," *Defense Update*, March 27, 2012a. As of August 8, 2013:
http://defense-update.com/20120327_israeli-multi-purpose-tank-ammo-redesigned-to-fit-the-125mm-gun-of-the-t-90s.html

———, "Elbit Systems Extends Dismounted C2 System to Infantry Squads and SF Teams," *Defense Update*, October 18, 2012b, As of June 24, 2014:
http://defense-update.com/20121018_dominator_ld.html

"Eurocopter 665 Tiger/Tigre," *Jane's Defence Equipment and Technology*, May 13, 2013.

"Eurocopter Tiger," *Military-Today.com*, undated As of June 23, 2014:
http://www.military-today.com/helicopters/eurocopter_tiger.htm

"Executive Overview: Logistics Support and Unmanned Vehicle Technology," *Land Warfare Platforms: Logistics, Support & Unmanned*, May 20, 2014. As of July 9, 2014:
https://janes.ihs.com/CustomPages/Janes/DisplayPage.aspx?DocType=Reference&ItemId=+++1508278&Pubabbrev=JLSU

"Fact Sheet: Research and Development of Tarian," Defence Science & Technology Laboratory, United Kingdom Ministry of Defence, 2012.

"Fajr 5," *Global Military Review*, undated. As of June 23, 2014:
http://globalmilitaryreview.blogspot.com/2012/09/iranian-fajr-5-multiple-launch-rocket.html

"FELIN (Fantassin à Équipements et Liaisons Intégrés)—Future Infantry Soldier System, France," *Army-Technology.com*, undated. As of June 23, 2014:
http://www.army-technology.com/projects/felin/

Fiorenza, Nicholas, "Rheinmetall Give Them HEL," *Aviation Week*, October 28, 2013. As of June 23, 2014:
http://aviationweek.com/blog/rheinmetall-gives-them-hel

Fish, Tim, "QinetiQ's PACSCAT Demonstrator Chosen for FLC Trials," *Jane's Defence Weekly*, August 18, 2010.

"FIST—Future Infantry Soldier Technology, United Kingdom," *Army-Technology.com*, undated. As of June 23, 2014:
http://www.army-technology.com/projects/fist/

"Flight of the Drones: Why the Future of Air Power Belongs to Unmanned Systems," *The Economist*, October 8, 2011. As of June 23, 2014:
http://www.economist.com/node/21531433

Foss, Christopher F., "China Looks to Export Air-Transportable IFV," *Jane's International Defense Review*, July 6, 2012a.

———, "France Seeks Out Improved Artillery Projectiles," *Jane's International Defense Review*, July 9, 2012b.

———, "International Armored Support Vehicles," *Jane's International Defence Review*, 2012c.

———, "Skyshield Can Fire AHEAD," *Israeli Homeland Security*, February 20, 2013. As of June 23, 2014:
http://www.ihs.com/events/exhibitions/idex-2013/news/feb-20/Skyshield-can-fire-far-AHEAD.aspx

Freedberg, Sydney J., Jr., "Army Plays Shell Game with Unfinished Apache Helicopters: Put the Transmission in, and Pull It out Again," *Breakingdefense.com*, April 26, 2013. As of June 23, 2014:
http://breakingdefense.com/2013/04/26/army-plays-shell-game-with-unfinished-apache-helicopters-put-th/

Froysland, Jeff, and Scott Prochniak, "Training and Doctrine Command Capability Manager—Fires Brigade," *FIRES Journal*, March–April 2013, pp. 41–42.

"Future Soldier on the March: Rheinmetall Hands Over 'Gladius' Soldier System to the Bundeswehr," Presse Box, March 11, 2013. As of June 23, 2014:
http://www.pressebox.com/pressrelease/rheinmetall-ag/Future-Soldier-on-the-march-Rheinmetall-hands-over-Gladius-soldier-system-to-the-Bundeswehr/boxid/580156

"FV432 APC," *Military Factory*, last update July 7, 2013 As of June 23, 2014:
http://www.militaryfactory.com/armor/detail.asp?armor_id=165

"General Dynamics Land Systems—Force Protection Cougar Mine Resistant Ambush Protected Vehicle," *Jane's Defense & Security Analysis*, February 28, 2013.

"General Dynamics Land Systems M1/M1A1/M1A2 Abrams MBT," *Jane's Armor and Artillery*, last updated April 3, 2012.

Gourley, Scott, "Aviation Modernization Milestone Update," *Army Magazine*, January 2013. As of June 23, 2014:
http://www.ausa.org/publications/armymagazine/archive/2013/01/Documents/GourleyAviation_January2013.pdf

"GROM Universal Fighting Module," *Kharkiv Morozov Machine Building's Design Bureau*, undated. As of August 6, 2013:
http://www.morozov.com.ua/eng/body/grom.php

Grunden, Hunter, "Special Operators Welcome Small Drones, but Need Better Sensors," *National Defense Magazine*, November 2006. As of June 23, 2014: http://www.nationaldefensemagazine.org/archive/2006/November/Pages/SpecialOperators2805.aspx

"Guardium UGV G-NIUS Semi-Autonomous Unmanned Ground Systems," *Army Recognition*, undated. As of June 23, 2014: http://www.armyrecognition.com/israel_israeli_wheeled_armoured_and_vehicle_uk/guardium_ugv_semi-autonomous_unmanned_ground_system_vehicle_g-nius_israeli_army_israel_pictures_tech.html

H-1 Program: AH-1Z and UH-1Y, Naval Air Systems Command, PMA-276 pamphlet, 2013.

"Honeywell Unmanned Aerial Vehicle Included in Army's Contract for Brigade Combat Team Modernization Increment 1," press release, Honeywell, August 24, 2010. As of June 23, 2014: http://www51.honeywell.com/honeywell/news-events/press-releases-details/8.24.10THawkIncrement1.html

Hornyak, Timothy, "Korean Machine-Gunrobots Start DMZ Duty," *CNET*, July 14, 2010. As of June 23, 2014: http://news.cnet.com/8301-17938_105-20010533-1.html

Huls, Harlan, "Firing US 120mm Tank Ammunition in the Leopard 2 Main Battle Tank," presented at the NDIA Guns and Missiles Conference, April 22, 2008. As of June 23, 2014: http://www.dtic.mil/ndia/2008gun_missile/6526Huls.pdf

"IMI Israel Military Industries Presents the LYNX Multi-Caliber Rocket Launcher System at MSPO 2013," *Army Recognition*, September 3, 2013. As of June 25, 2014: http://www.armyrecognition.com/mspo_2013_show_daily_news_coverage_report/imi_israel_military_industries_presents_lynx_multicaliber_rocket_launcher_system_mspo_2013_0309133.html

"IMI to Present Products and Capabilities Suitable for the Colombian Market at F-AIR 2013," *Defense-Aerospace.com*, July 2013.

"India Seeks M777 155mm Howitzers from US," *Army-Technology.com*, August 9, 2013. As of June 23, 2014: http://www.army-technology.com/news/newsindia-seeks-m777-155mm-howitzers-from-us

"Iran Unveils Indigenous Stealth Reconnaissance, Combat Drone," *Press TV*, May 9, 2013. As of June 23, 2014: http://www.presstv.com/detail/2013/05/09/302684/iran-unveils-stealth-recce-combat-drone

"Israel Military Industries Bulldozer Protection Kit," *Jane's Defense & Security Analysis*, July 13, 2012.

"ISW Tytan System," *MyCity-Military*, undated. As of June 23, 2014:
http://www.mycity-military.com/Pesadijsko-naoruzanje-municija-i-oprema/Komandno-komunikacioni-uredjaji_20.html

"Italian 76mm *Draco* System," *Military-Today.com*, undated. As of July 9, 2014:
http://www.military-today.com/artillery/draco.htm

Jensen, David, "What's New with CAAS?" *Rotor & Wing Magazine*, October 1, 2010. As of June 23, 2014:
http://www.aviationtoday.com/rw/issue/features/71009.html

"K9 Thunder," *Army Guide*, undated. As of June 23, 2014:
http://www.army-guide.com/eng/product1279.html

Karcher, Timothy, "Enhancing Combat Effectiveness, the Evolution of the United States Army Infantry Rifle Squad Since the End of World War II," master's thesis, United States Army Command and General Staff College at Fort Leavenworth, Kan., 2002.

"Katushya Rocket," *GlobalSecurity.org*, undated. As of June 23, 2014:
www.globalsecurity.org/military/world/russia/katyusha.htm

Keller, John, "Boeing Moves Apache Block III Attack Helicopter Program Forward with $187 Million Army Contract," *MilitaryAerospace.com*, March 18, 2012a. As of June 23, 2014:
www.militaryaerospace.com/articles/2012/03/apache-block-iii.html

———, "JIEDDO Seeks to Shrink Soldier-Worn IED Detector Technology to Less Than 20 Pounds," *Military and Aerospace Electronics*, October 18, 2012b. As of June 23, 2014:
http://www.militaryaerospace.com/articles/2012/10/20-pound-ied-detector.html

"Kharkov Morozov Design Bureau BTR-4 Armoured Personnel Carrier," *Jane's Land Warfare Platforms: Armoured Fighting Vehicles*," last updated January 26, 2012.

Kopp, Carlo, *Russian and PLA Point Defence System Vehicles*, *Air Power Australia*, June 2008, updated April 2012 As of June 24, 2014:
http://www.ausairpower.net/APA-PD-SAM-SPAAG-TEL-TL.html

"Korea's Future Warrior," *Soldier Modernisation*, Vol. 7, June 2011. As of June 23, 2014:
http://www.soldiermod.com/volume-7/korea.html

"Krassu-Maffei Wegmann Pazerhaubitze 2000 (PzH 2000)," *Jane's Armour and Artillery*, updated February 7, 2012.

"Landing Craft Utility (LCU)," GlobalSecurity.org, undated. As of July 18, 2014:
http://www.globalsecurity.org/military/systems/ship/lcu-2000.htm

"Leopard 2A6," *Defense Update*, 2007. As of June 23, 2014:
http://defense-update.com/newscast/0407/news/160407_leo2.htm

Lesiak, Brian, email correspondence, April 1, 2013.

Levy-Stein, Revital, and Gili Cohen, "Iron Dome Battery Successfully Intercepts Target," Haaretz, August 13, 2008. As of June 23, 2014:
www.haaretz.com/news/diplomacy-defense/.premium-1.541152

"Light Armoured Vehicle (LAV) III," Canadian Army, modified October 23, 2013. As of June 23, 2014:
http://www.army-armee.forces.gc.ca/en/vehicles/light-armoured-vehicle.page

"Litter Evacuation," Chapter Nine in *Medical Evacuation in a Theater of Operation Tactics, Techniques, and Procedures*, Field Manual 8-10-6, Washington, D.C.: Headquarters, Department of the Army, April 14, 2000. As of June 23, 2014:
http://mediccom.org/public/tadmat/training/NDMS/Litter_evac_1.pdf

"Lockheed Martin Missiles and Fire Control 227 mm Multiple Launch Rocket System (MLRS)," *Jane's Armour and Artillery*, updated July 25, 2013.

"Lockheed Martin Missiles and Fire Control M142 227 mm (6-Round) High-Mobility Artillery Rocket System (HIMARS)," *Jane's Armour and Artillery*, updated July 25, 2013.

"Logistics Knights in Shining Armour," *Jane's Defence Weekly*, May 11, 2010.

"Lynx Autonomous Multi-Purpose Rocket System," Israeli Military Industries, undated. As of August 5, 2013:
http://www.imi-israel.com/vault/documents/lynx_eng.pdf

"M113 AS4 Armoured Personnel Carrier," Australian Army, undated. As of June 23, 2014:
http://www.army.gov.au/Our-work/Equipment-and-clothing/Vehicles/M113AS4

"M136 AT4," *Military Analysis Network*, updated January 8, 1999. As of June 23, 2014:
http://www.fas.org/man/dod-101/sys/land/at4.htm

"M198 Howitzer," *Military.com*, undated. As of June 23, 2014:
http://www.military.com/equipment/m198-howitzer

"M1A2 Abrams," *Defense Industry Daily*, February 4, 2014. As of June 23, 2014:
http://www.defenseindustrydaily.com/digital-abrams-the-m1a2-sep-program-updated-02834/

"M270 MLRS," Federation of American Scientists, February 22, 2000. As of June 23, 2014:
https://www.fas.org/man/dod-101/sys/land/m270.htm

Majumdar, Dave, "US Army Fields First AH-64E Unit, but More Improvements to Come," *FlightGlobal.com*, January 9, 2013. As of June 23, 2014:
http://www.flightglobal.com/news/articles/us-army-fields-first-ah-64e-unit-but-more-improvements-to-come-380875

Maneuver Center of Excellence personnel, interviews, Ft. Benning, Columbus, Ga., July 2013.

"Marines Up Order for New Heavy Lifter," *Rotor & Wing*, August 1, 2007. As of June 23, 2014:
http://www.aviationtoday.com/rw/military/heavylift/14482.html

"Mass Production of Mi-26T2 Will Begin in 2012 Year," *Aviationunion.org*, January 3, 2012. As of January 3, 2012:
http://www.aviationunion.org/news_second.php?new=99

Matsumura, John, Randall Steeb, Matthew W. Lewis, Kathryn Connor, Timothy Vail, Matthew E. Boyer, Steve DiAntonio, Jan Osburg, Kristy Gonzalez Morganti, James McGee, and Paul S. Steinberg, *Assessing the Impact of Autonomous Robotic Systems on Army Force Structure*, Santa Monica, Calif.: RAND Corporation, RR-226-A, 2012.

McIlvaine, Rob, "Squad Needs 'Overmatch' Capability," U.S. Army, October 13, 2011. As of July 9, 2014:
http://www.army.mil/article/67175/

"Merkava Mk 4 MBT," *Jane's Armor and Artillery*, last updated March 28, 2012.

"Merkava Mk 4 Min Battle Tank, Israel," *Army-Technology.com*, undated. As of June 23, 2014:
http://www.army-technology.com/projects/merkava4/merkava41.html

"Mi-26T2," *Deagel.com*, August 11, 2011. As of June 23, 2014:
http://www.deagel.com/Tactical-Support-Helicopters/Mi-26T2_a000359002.aspx

"Mi-26 T2: Multipurpose Transport Helicopter," *Redstar.gr*, undated. As of June 23, 2014:
http://www.redstar.gr/Foto_red/Eng/Helicopter/Mi26T2.html

"Mi-26T2 Versus CHINOOK," *Take-off Magazine*, February 2011, pp. 16–18. As of June 23, 2014:
http://en.take-off.ru/pdf_to/to19.pdf

"Mi-28A/N Havoc Attack Helicopter, Russian Federation," *Army-Technology.com*, undated. As of June 23, 2014:
http://www.army-technology.com/projects/mi28/

"Mi-28 Russian Havoc," *Worldwide-Military*, undated. As of June 23, 2014:
http://www.worldwide-military.com/Military%20Heli's/Attack%20heli/Links/Mi-28_Havoc_EN.htm

"Mi-171A2: Another Step Forward," *Russianhelicopters.com*, 2012. As of June 23, 2014:
http://tangentlink.com/wp-content/uploads/2013/09/5.-Russian-Helicopters-Experience-Innovation-Nikolay-Titov.pdf

"Mi-171A2 New Medium Multirole Helicopter," *Russianhelicopters.com*, undated. As of June 23, 2014:
http://www.russianhelicopters.aero/en/helicopters/civil/mi-171a2.html

"Mi-171M: New Life of Venerable Helicopter," *Take-off Magazine*, July 2010, pp. 12–13. As of June 23, 2014:
http://en.take-off.ru/pdf_to/to17.pdf

"Mil Mi-17 (Mi-8M), Mi-19, Mi-171 and Mi-172," *Jane's Defence Equipment and Technology*, February 7, 2013. As of June 23, 2014:
https://janes.ihs.com/CustomPages/Janes/DisplayPage.aspx?DocType=Reference&ItemId=+++1342930&Pubabbrev=JAWA

"Mil Mi-26 and Mi-27," *Jane's Defence Equipment and Technology*, June 24, 2013. As of June 23, 2014:
https://janes.ihs.com/CustomPages/Janes/DisplayPage.aspx?DocType=Reference&ItemId=+++1311619&Pubabbrev=JHMS

"Mil Mi-28," *Jane's All the World's Aircraft*, last posted January 31, 2014. As of July 9, 2014:
https://janes.ihs.com/CustomPages/Janes/DisplayPage.aspx?DocType=Reference&ItemId=+++1342933&Pubabbrev=JAWA

"Mil Mi-28 Havoc Attack Helicopter," *Military-Today.com*, undated. As of June 23, 2014:
http://www.military-today.com/helicopters/mil_mi28_havoc.htm

"The Mil Mi-28 Havoc Has Since Become the Standard Attack Helicopter for the Russian Air Force and Army," *MilitaryFactory.com*, last updated February 26, 2014. As of July 9, 2014:
http://www.militaryfactory.com/aircraft/detail.asp?aircraft_id=156

Mizroch, Amir, "Israel's Rocket-Intercepting Ace Who Started Out on Warcraft," *Wired U.K.*, November 19, 2012. As of June 23, 2014:
http://www.wired.co.uk/news/archive/2012-11/19/israeli-rocket-hunting-ace

"Modern PLA Armoured Vehicles," *Air Power Australia*, undated. As of June 23, 2014:
http://www.army-guide.com/eng/product237.html

"Moving Forwards," *Soldier Modernisation*, Summer 2008. As of June 23, 2014:
http://www.soldiermod.com/summer-08/prog-mer.html

"Naval Construction, Naval Forecast and Naval Upgrades," Carpenter Data Partnership (CDP) Group, 2011. As of May 5, 2013:
http://www.carpenterdata.net/naval_construction.html

A New Equipping Strategy: Modernizing the U.S. Army of 2020, national security report, torchbearer issue, Arlington, Va.: Institute of Land Warfare and Association of the Untied States Army, June 2012.

"New Heavy Lift Helicopter Starts Development," U.S. Marine Corps, press release, January 9, 2006. As of June 23, 2014:
http://www.defense.gov/transformation/articles/2006-01/ta010906a.html

"New Mi-34C1, Ka-226T, Mi-38, Mi-26T2 Showcased at MAKS 2011," *Russianhelicopter.aero.en*, August 6, 2011. As of June 23, 2014:
http://www.russianhelicopters.aero/en/press/news/1870.html

"Nexter Systems CAESAR 155 mm Self-Propelled Gun," *Jane's Armour and Artillery*, updated February 7, 2012.

"NORICUM GH N-45 155 mm Gun-Howitzer," *Jane's Armour and Artillery*, updated March 1, 2012.

"NORINCO 155 mm Self-Propelled Gun Howitzer PLZ-05," *Jane's Armour and Artillery*, updated February 7, 2012.

"NORMANS Trials Update," *Soldier Modernisation*, Vol. 5, May 2010. As of June 23, 2014:
http://www.soldiermod.com/volume-5/normans.html

"Norwegians Choose Thales for NORMANS Soldier System," *Soldier Systems*, undated. As of June 23, 2014:
http://soldiersystems.net/2011/10/14/norwegians-choose-thales-for-normans-soldier-system/

Oestergaard, Joakim Kasper, "About the CH-47 and MH-47," last updated May 19, 2014. As of July 9, 2014
http://www.bga-aeroweb.com/Defense/CH-47-Chinook.html

Osborn, Kris, "Technology Gives Apache Block III More Lift, Capability, Landing Ability," Army.mil, February 26, 2010. As of June 23, 2014:
http://www.army.mil/article/35056/technology-gives-apache-block-iii-more-lift-capability-landing-ability/

Osborn, Kris, "Army Developing New Self Propelled Howitzer," *Army.mil*, September 1, 2011. As of June 24, 2014:
http://www.army.mil/article/64728/

Osborne, Anthony, "Boeing Hopeful of Indian Contracts Progress," *Aviation Week*, February 13, 2014. As of June 23, 2014:
http://aviationweek.com/defense/boeing-hopeful-indian-contracts-progress

"Pantsyr S1 Close Range Air Defence System, Russia," *Army-Technology.com*, undated. As of June 23, 2014:
www.army-technology.com/projects/pantsyr/

"Patria Land Systems Armoured Modular Vehicle," *Jane's Armour and Artillery*, last updated November 30, 2011.

"Phalanx Close-In Weapon System," video, *Military.com*, posted by "GunFun," June 15, 2012. As of June 26, 2014:
http://www.military.com/video/guns/naval-guns/phalanx-close-in-weapon-system/1691590028001/

"Photo: Iran's Brand New Drones (Including an Israeli UAV Clone) Exposed During Recent Wargames," *The Aviationist*, July 4, 2012. As of June 23, 2014:
http://theaviationist.com/2012/07/04/new-iranian-uav/

"Pioneer," *Israeli-Weapons.com*, undated. As of June 23, 2014:
http://www.israeli-weapons.com/weapons/aircraft/uav/pioneer/Pioneer.html

"PzH 2000," *Army Guide*, undated. As of June 23, 2014:
http://www.army-guide.com/eng/product237.html

Randolph, Monique, "New Digs Equal New Opportunities for Gruntworks Team," Marine Corps Systems Command, *MarCorSysCom.Marines.mil*, March 19, 2013. As of June 24, 2014:
http://www.marcorsyscom.marines.mil/News/PressReleaseArticleDisplay/tabid/8007/Article/139826/new-digs-equal-new-opportunities-for-gruntworks-team.aspx

Reed, John, "Army Fielding Robo Jeeps in A'Stan," *DefenseTech.org*, August 5, 2011. As of June 23, 2014:
http://defensetech.org/2011/08/05/army-fielding-robo-jeeps-to-astan/

"Rheinmetall Defence Wins New Contracts for Air Defence Systems from Malaysia and Kuwait," *Army Recognition*, January 23, 2013. As of June 23, 2004:
http://www.armyrecognition.com/january_2013_army_military_defense_industry_news/rheinmetall_defence_wins_new_contracts_for_air_defence_systems_from_malaysia_and_kuwait_2101134.html

"Rheinmetall Demonstrates 50kw HEL Laser," *Optics.org*, December 19, 2012. As of June 23, 2014:
http://optics.org/news/3/12/31

"Rheinmetall Landsysteme Condor," *Military Factory*, last updated June 24, 2013. As of June 23, 2014:
http://www.militaryfactory.com/armor/detail.asp?armor_id=516

"Rheinmetall Starts Gladius Soldier System Deliveries to German Army," *Army-Technology.com*, March 14, 2013. As of June 23, 2014:
http://www.army-technology.com/news/newsrheinmetall-starts-gladius-soldier-system-deliveries-german-army

"RIM-116 Rolling Airframe Missile," United States Navy Fact File, *Navy.mil*, last updated November 19, 2013. As of June 24, 2014:
http://www.navy.mil/navydata/fact_display.asp?cid=2200&tid=800&ct=2

Ripley, Tim, "UK Awards Contract for Wolfhound Mine-Protected Supply Vehicles," *Jane's Defence Weekly*, April 9, 2009. As of July 9, 2014:
https://janes.ihs.com/CustomPages/Janes/DisplayPage.aspx?DocType=News&ItemId=+++1179932&Pubabbrev=JDW

"Rolling Airframe Missile (RAM) Guided Missile System," *Raytheon*, undated. As of June 23, 2014:
http://www.raytheon.com/capabilities/products/ram/

"Rostvertol Will Demonstrate the Modernized Mi-26T2 Heavy Transport Helicopter to Algerian Air Forces," *RussianAviation.com*, June 20, 2012. As of June 23, 2014:
http://www.ruaviation.com/news/2012/6/20/1052/

"RQ-16: Future Combat Systems' Last UAV Survivor Falls," *Defense Industry Daily*, September 19, 2012. As of June 23, 2014:
http://www.defenseindustrydaily.com/one-small-step-for-a-uav-one-big-step-for-fcs-class-i-01372/

"Russian Air Force Receives First Mi-28 Night Hunter Helicopter," *RiaNovosti*, June 5, 2006. As of June 23, 2014:
http://en.rian.ru/russia/20060605/49063536.html

"Russian Army Postpones Future Soldier Uniform Induction," *Army-Technology.com*, July 15, 2013. As of June 23, 2014:
http://www.army-technology.com/news/newsrussian-army-postpones-future-soldier-uniform-induction

"Russian Future Soldier Ratnik Will Be Demonstrated at Defense Exhibition REA 2013 in September," *Army Recognition*, July 12, 2013. As of June 23, 2014:
http://www.armyrecognition.com/july_2013_news_defence_security_industry_military/russian_future_soldier_ratnik_will_be_demonstrated_at_defense_exhibition_rea_2013_september_1207131.html

"Russian Helicopters Delivers First Mi-171A2 Fuselage," *RiaNovosti*, January 23, 2012. As of June 23, 2014:
http://en.rian.ru/russia/20120123/170902468.html

"Russian Mi-171A2," *Russian Helicopters*, undated. As of June 23, 2014:
www.russianhelicopters.aero

"Russian Military to Purchase 10-15 Mi-28N Helicopters per Year," *RiaNovosti*, January 22, 2008. As of June 23, 2014:
http://en.rian.ru/russia/20080122/97530653.html

"Russians Inspect Soldato Futuro," *Soldier Systems*, undated. As of June 23, 2014:
http://soldiersystems.net/2012/03/14/russians-inspect-soldato-futuro/

"Russia to Commission BMD-4M Airborne Vehicles in 2013," *RiaNovosti,* December 27, 2012.

Rustechnologies, "Russia Presents Mi-171A2 Helicopter with New Avionics," *Ros Technologies Blog*, June 19, 2013. As of June 23, 2014:
http://rostechnologiesblog.wordpress.com/2013/06/19/russia-presents-mi-171a2-helicopter-with-new-avionics/

"A Safe Place for Medicine," *Army-Technology*, undated. As of June 23, 2014:
http://www.army-technology.com/features/feature87779/feature87779-7.html

"Samsung Techwin 155 mm/52-Calibre K9 Thunder Self-Propelled Artillery System," *Jane's Armour and Artillery*, updated February 7, 2012.

Sattler, V., and M. O'Leary, "Organizing Modern Infantry: An Analysis of Section Fighting Power," *Canada Army Journal*, Vol. 13, No. 3, Autumn 2010, pp. 23–53.

"SCAIC 400 mm WS-2 Multiple Rocket Weapon System," *Jane's Armour and Artillery*, updated July 22, 2013.

Scott, Richard, "BMT Ramps up Tri-Bow Monohull Landing Craft Concept," *International Defence Review*, April 12, 2010.

———, "Race to the Beach: Novel Hullforms Push the Pace," *Jane's Navy International*, April 20, 2011.

"Sikorsky Aircraft Selects Rockwell Collins to Provide CH-53K Avionics Management System," *Sikorsky.com*, June 29, 2006.

Sikorsky CH-53K Helicopter, brochure, *Sikorsky.com*, June 2007.

"Sikorsky CH-53K Super Stallion," *Jane's Defence Equipment and Technology*, August 3, 2012.

"Sikorsky S-70 (H-60) Upgrades," *Jane's Defence Equipment and Technology*, last posted January 23, 2013. As of June 23, 2014:
https://janes.ihs.com/CustomPages/Janes/DisplayPage.aspx?DocType=Reference&ItemId=+++1337691&Pubabbrev=JAU_

"Sikorsky S-70A," *Jane's Defence Equipment and Technology*, February 7, 2013. As of June 23, 2014:
https://janes.ihs.com/CustomPages/Janes/DisplayPage.aspx?DocType=Reference&ItemId=+++1343474&Pubabbrev=JAWA

Sikorsky UH-60M Black Hawk Helicopter, brochure, *Sikorsky.com*, July 2009.

Singh, Ranjeet, "Active Protection Systems," *South Asia Defence and Strategic Review*, July 26, 2011. As of August 8, 2013:
http://www.defstrat.com/exec/frmArticleDetails.aspx?DID=306

"Soldato Futuro Future Soldier System, Italy," *Army-Technology.com*, undated As of June 23, 2014:
http://www.army-technology.com/projects/italiansoldiersystem/

"A Soldier's Burden," *Jane's Defence Weekly*, last posted March 7, 2013. As of June 23, 2014:
https://janes.ihs.com/CustomPages/Janes/DisplayPage.aspx?DocType=News&ItemId=+++1548463&Pubabbrev=JDW

"Soldier Systems Modernization in the Norwegian Armed Forces," *Soldier Modernisation*, Vol. 7, June 2011. As of June 23, 2014:
http://www.soldiermod.com/volume-7/normans.html

"SPLAV 122mm BM-21 Multiple Rocket Launcher Family," *Jane's Armour and Artillery*, updated July 23, 2013.

"SPLAV 300 mm BM 9A52 (12-Round) Smerch Multiple Rocket System," *Jane's Armour and Artillery*, updated July 23, 2013.

Stoker, Liam, "January's Top Stories: More Redundancies and Army Modernization," *Army-Technology.com*, February 1, 2013. As of June 23, 2014:
www.army-technology.com/features/featurejanuary-top-stories-redundancies-army-modernisation

"Stryker," U.S. Army Fact Files, *Army.mil*, undated. As of July 31, 2013:
http://www.army.mil/factfiles/equipment/wheeled/stryker.html

"Sweden to Field Air Burst Grenade in 2011," *Soldier Modernisation*, Vol. 6, January 2011. As of June 23, 2014:
http://www.soldiermod.com/volume-6/grenade.html

"T-90S Main Battle Tank, Russia," *Army-Technology.com*, undated. As of June 23, 2014:
http://www.army-technology.com/projects/t90/

"Tiger Multi-Role Combat Helicopter, Germany," *Army-Technology.com*, undated. As of June 24, 2014:
http://www.army-technology.com/projects/tiger/

"TopGun," *IAI.com*, undated. As of August 14, 2013:
http://www.iai.co.il/16147-44363-en/Business_Areas_Military_Land_PrecisionStrike.aspx?btl=1

Trimble, Stephen, "From Albania to Afghanistan, US Army Integrates Lessons into Latest Apache," *FlightGlobal.com*, November 3, 2011. As of June 24, 2014:
http://www.flightglobal.com/news/articles/from-albania-to-afghanistan-us-army-integrates-lessons-into-latest-apache-364245

Tusa, Francis, "Wagon Train: Logistics Lessons from Operations," *Jane's International Defence Review*, November 7, 2012.

"Type 69 40 mm Airburst Anti-Personnel Round," *Jane's Infantry Weapons*, last posted August 19, 2010. As of June 23, 2014:
https://janes.ihs.com/CustomPages/Janes/DisplayPage.aspx?DocType=Reference&ItemId=+++1362725

"Tytan at a Turning Point," *Soldier Modernisation*, Vol. 3, June 2009. As of June 23, 2014:
http://soldiermod.com/volume-2-06/tytan.html

"UAE Order 40 K11 Airburst Rifles from South Korea," *Asian Defence*, 2010. As of June 23, 2014:
http://theasiandefence.blogspot.com/2010/05/uae-order-40-k11-airburst-rifles-from.html

"UH-60M," *Globalsecurity.com*, last modified July 7, 2011. As of June 23, 2014:
http://www.globalsecurity.org/military/systems/aircraft/uh-60m.htm

"UH-60M Black Hawk Helicopter," *Sikorsky.com*, undated.

"UH-60M Evolution" PowerPoint presentation, undated. As of June 23, 2014:
http://www.hawkpilot.com/UH60Info/Obrion%20Classes/UH-60M%20Evolution.ppt

"Unlike Other Attack Helicopters in Its Class, the Eurocopter Tiger Sits the Pilot in the Front Cockpit and the Weapons Officer in the Rear Cockpit," *MilitaryFactory.com*, June 8, 2011.

U.S. Congress, 106th Cong., National Defense Authorization, Fiscal Year 2001, Public Law 106-398, Washington, D.C., October 30, 2000.

U.S. Government Accountability Office, *Defense Acquisitions: Assessments of Selected Weapon Programs*, GAO-13-294SP, March 2013. As of June 23, 2014:
http://www.gao.gov/assets/660/653379.pdf

"USMC Rifle Squad and Irrigation Ditch," *SinoDefenceForum.com*, posted by "bd popeye," July 7, 2012. As of June 23, 2014:
http://www.sinodefenceforum.com/world-military-pictures/us-military-pictures-thread-88-1975.html

"The USMC's Expeditionary Fighting Vehicle (EFV)," *Defense Industry Daily*, June 26, 2012. As of June 23, 2014:
http://www.defenseindustrydaily.com/the-usmcs-expeditionary-fighting-vehicle-sdd-phase-updated-02302/

Valpolini, Paolo, "The Weight of Intelligence," *Armada International*, April 1, 2010. As of June 23, 2014:
http://www.readperiodicals.com/201004/2036413171.html

———, "Italy's Soldato Futuro Moves Towards Production," *Soldier Modernisation*, Vol. 8, January 2012. As of June 23, 2014:
http://www.soldiermod.com/volume-8/soldato-futuro.html

Warwick, Graham, "Block 2 CH-47F to Tackle Payload Shortfalls," *Military.com*, April 22, 2013. As of June 23, 2014:
http://www.military.com/daily-news/2013/04/22/block-2-ch47f-to-tackle-payload-shortfalls.html

Weatherington, Dyke D., Director, Unmanned Warfare & Intelligence, Surveillance, and Reconnaissance, personal communication, telephone, 2013.

Williams, Huw, "Future Protected Vehicle Study Turns up Host of Concepts and New Technologies," *Jane's International Defence Review*, January 11, 2001.

Wilson, J. R., "Sea Soldier's Load," *DefenseMediaNetwork.com*, November 9, 2010. As of June 24, 2014:
http://www.defensemedianetwork.com/stories/sea-soldiers-load-the-marines-personal-gear-today/

"WS-2 Multiple Launch Rocket System," *Military-Today.com*, undated. As of June 23, 2014:
http://www.military-today.com/artillery/ws2.htm

"Z-10 Attack Helicopter, China," *Army-Technology.com*, undated. As of June 23, 2014:
http://www.army-technology.com/projects/z-10-attack-helicopter-china-liberation-army/

(This page is intentionally blank.)